WINTER

January

The most powerful thing God uses in my life to teach me, to discipline me, and to love me with is the earth. Celebrate her goodness, and you shall feel the breasts of God.

—Elizabeth Anna Samudio

OVERVIEW

January is a good time to take stock. All of the warm season plants have either died back or spent their lives to return to the earth. Winter's starkness is an invitation to observe the bare essentials of your outdoor space. Perhaps, in the summer with all the over growth you felt overwhelmed by clutter, well now is your moment to recapture your planting areas, make new plans, and day dream about spring, which is just around the corner.

JANUARY CHECKLIST

- ☐ Observe
- ☐ Order seeds
- ☐ Start composting

1. OBSERVATION

Observation is a lost rhythm among the not-so child at heart. If you have not made a good New Years resolution, this may be the one you need – to wonder, reflect, and give time to the art of observation.

If you do not know where to start, observe children, the ones that prefer the outdoors to electronics. Their inclination to curiosity keeps them from getting bored and drives them to great discoveries, and if you mimic that behavior in your gardening quest, you too will become as a child and enter the plant and animal kingdom!

You will need to get to know your site, your life, and an estimate on the time it takes to bridge them together so that you can steward an urban farm or small garden.

The Unconventional Edible Garden

GROWING FOOD IN TEXAS
AND OTHER HARD AND DIFFICULT PLACES

by ELIZABETH ANNA SAMUDIO

a **Permaculture Press** publication

Dedications

All I ever wanted to do was make the world a better place, so I grew food in spaces close to people.

As a little girl, my grandfather gave me a love for ranching and an eye for natural systems; my grandmother, a love for the vegetable garden and preservation of food; my mother, an appreciation of the ornamental garden and a passion for beauty; my father, a love of developing spaces with heavy equipment; my older brother, Jaime, my boisterous confidence which he gave me when he dared me to take the first swing on the old rope that hung in our grandparent's barn and I soared over the hay, where I would either drop by what at the time seemed like 100 feet, or come back and hit the make shift ladder; my younger brother, Jacob, good company and his like-hearted spirit, as he loved to spend hours walking with me on the logging trails covered with moss and fern, deep into the forest – but it was God who gave me a love of nature.

My family is my plow.

I dedicate this book to them: those that are here with me now and those whose memory I hold tight. To my husband, the love of my life, who through two cancers kept me alive with his faith and devotion. To my sons and grandchildren, who *all* love the farm, ranch, garden, and God's creation.

Special thanks to my granddaughter, Melody May, who cherishes her Nana's place and makes it her own, *Elizabeth Anna Urban Farm*.

Blessed to be part of a break-through heritage.

Elizabeth Anna Samudio

Also known as Rhonda Henifin

Contents

Foreword

Today, many of us live and contribute to a way of life has become so specialized we have lost our connection to our food. The convention of gardening has unintentionally been removed from our experience. What do the busy parents grabbing fast-prepared food and the guy pushing his plastic cart through a maze of supermarket offerings have in common? They both have strayed from their partnership with the earth. With some effort and piles of joy, they can reconnect and make the unconventional a common and productive part of our lives.

My sister Elizabeth, or Rhonda as I know her, has never lost her connection to the earth through the soil, light, water, and creation. I can still taste the snap peas she picked for little baby Jake from my grandma's labyrinthine garden in Eastern Washington (which has summer weather and temperatures comparable to a cool Texas summer). Dry, dusty soil gave way to cool mud pushing through our toes as we sauntered through the vegetable furrows. To the five-year-old me, the humid, sweetly pungent garden was a jungle of food and life. With butterflies, bees, and the occasional newt, it was a wonderland to explore. Rhonda taught me more about Grandma Betty's vegetables than I can remember to this day, and I have often returned to those early days when I turned the soil barefoot with shovels and hands...

Elizabeth is taking this knowledge and passion to teach us how to connect with the soil in difficult conditions. With her sweet and insightful voice, she shares with us how to turn the hardscrabble into fertile grounds to grow our own food. At the same time, we learn and remember how vital the soil is to our health and mental well-being. We are further reminded that in simple, thoughtful terms, we can reconnect to and recharge by making what was once conventional (growing our own food), conventional once again.

baby jake, 2016

JACOB HENIFIN
Elizabeth's younger brother

To my auntie, so kind and dear
A woman who brings about fun and cheer.

She is a lover of flowers and all things nature.
She is a believer in God, our greatest creator.

Everything my aunt touches feels her blessing
Love from her is never something to be guessing.

With a big smile she warms up my heart
And have you seen her style? She is walking art!

Bright colors is what she chooses in fashion
Teaching others to farm is what she chooses for passion

If you are feeling darkness and covered in gloom
Just listen to her laugh it will brighten a room.

She is my aunt, the one who chose a new name
Rhonda to Elizabeth – both beautiful I shall proclaim!

She walks to the beat of her own drum
Barefoot, silly and sweet as a plum.

So dear Aunt - I love you, your beauty and the fun times we share
But most of all I love the way you always show others you care!

JESSICA HENIFIN BREARLY
daughter of Elizabeth's older brother, Jaime

8

My First Garden

My first real garden I grew – and grew in – was an urban garden in Bellingham, a rural town in Washington state.

Though I had tilled, planted, harvested, and canned from a garden larger then the size of most back yards in a home I rented as well as several plots I carved out in college, I will always think of my garden on Patton Street as my first love.

It was there in an older neighborhood – with homes made of real wood siding and walking distance to Cornwall Park, a corner convenience store, a gravel pit, and another large urban family farm with lots of tomatoes, dahlias, and tall rows of sweet corn my boys ran through – that I learned about soil preparation.

At age 23 I took up my path. The virgin ground in the back corner of my yard, which was three-fourths of the land, called out for homegrown tomatoes, corn, sunflowers, carrots, and winter squash. Though in the northwest, this 300-square-foot area was still compacted, laden with nutrient rich clay, and in need of much work. So I started my first compost pile and hauled in lots of free horse manure in the back of my VW bus.

Being the impetuous female I was/am, and the fact that I had two toddlers at my shirttails, I first had my husband build a fence to keep our boys from wandering away, then dug a big hole and began stuffing it with all the organic matter I could get my hands on. Even the few leftovers we had became part of our compost pile – a mimicked forest floor.

Yes, there are dos and don'ts to composting, but I never liked reading rules too closely. Slowly but surely, my potpourri of organic collectibles turned into something beautiful – composted earth. And this is what I used to grow my first real edible garden. And I will say – she was a glorious playground for my boys, the neighborhood children, and me.

Elizabeth and her granddaughter Melody with a baby duck

The first essential step is to spend time observing natural things, and to ask yourself questions, such as:

+ Where is the sun exposure on your site? Get a compass if you are unsure. Remember the sun rises in the east and sets in the west.

+ Where are the prevailing winds?

+ What is the condition of the earth? Dig in!

+ Wild life – what hangs out on my site?

+ The roof – is there opportunity for water collection?

+ What is your pet's out-door routine?

+ The neighbors – do I know them?

+ The history of this place – what is it?

+ Drainage – can it be used to my advantage? If not, what can I do to keep water on my site, yet away from my home's foundation? Do I have slopes to take advantage of passive water collection in the earth?

+ Access – how easily can I get to the places I want to develop?

+ Where is the best place for a compost pile?

+ What is the health of my existing plants? This may take until spring to know for sure.

+ How is my health? Perhaps as I restore the earth, I need to restore my body as well so that I am up for the task. Hint: this work brings life to your mind, body and soul.

+ The functions of the existing plants – are they beneficial or hindering?

+ The natural edges of my site – where are they? Where do things collect? Do I have a fence for edible vines to grow? Do I have plants for birds to take shelter? Is the edge a place I can plant that will ease my labor since it is functioning as a collector of water and topsoils, etc?

+ My driveway – is there a space for winter gardening where I could add a planting box since the ground holds heat on concreted drives?

+ The lateral space—how can I use this?

Your turn: did you think of something else? To be a successful urban farmer, you will need to observe often: even daily for a while. Be sure to keep a journal of your observations, as all of these observations will help you determine how to make your site sustainable and alive!

Cheers to your brand new year!

2. WHAT TO PLANT?

Bear in mind that there are many more species one can plant, but this book is founded on making you a successful gardener in North Texas or other difficult places, so I have narrowed the scope so your focus is clear to increase success!

+ English peas

+ Onion sets

+ Bare-root fruit trees

+ Plus cool-season crops, if you have access to cold frames

Soil preparation is key to growing healthy plants; all seasoned gardeners know that healthy soils cover a multitude of mistakes. This is especially true when planting in the harsh times of winter or summer. If you have not yet learned how to build good soils, hold off before you plant seeds and starts. It is helpful to understand that in North Texas, and now more regions due to climate change, hard and difficult soils and disrupted cool-to-warm temperatures in winter months need to be factored in when choosing what to plant. I have found a few tree and vine varieties that perform well even in these harsh conditions: Anna apple, redskin peach, Oriental pear, Mexican plum, pawpaw, all persimmons, bare-root blackberry vine starts, and grape vines (try to find the variety that is indigenous to your area). My favorite for hot, dry climates is the Cherokee blackberry vine. She is a thorny lady, but the fruit is worth the thorn.

The things you need to watch for with fruit tree varieties are the number of days of chill at 45 degrees or below required for a plant to set her fruit (also known as "chill days") and whether or not the tree is a self-pollinator. Often we have short chill days in hot climates, so save yourself the burden of growing something that needs 900 hours of chilling time. Stick with low-chill varieties, as these are the trees that will produce fruit in areas such as ours where winters are dappled with weather above 45 degrees.

There are a lot more fruit trees that deserve a place on your small or large site, but often these uncommon varieties are not available as bare-root. My suggestion is that you keep some space for those other fabulous trees that are not as common, such as pineapple guava, pawpaw, goji bush, and jackfruit. More information about these gems and where to find them is covered in the October section, since fall is also a good time to plant potted fruit-bearing trees. This is not only true of Texas, but also most climates where fall brings forth new flora growth in the non-domestic landscape.

3. ORDERING SEEDS

Before the days of Google and social media, most gardeners subscribed to seed catalogues: Ferry Seeds, Woods Seeds, Morse's, and Burpee Seeds. Gray winter days were warmed by hearth fires and ottomans cluttered with stacks of lovely botanical publications. Established companies still provide printed catalogues, but they also have online catalogues to create less paper waste. It makes the most sense to buy online, unless you plan on keeping these around as source books.

Make sure the seeds you gather are not GMOs (Genetically Modified Organisms). In warmer urban climates, use simple cold frame techniques to plant early crops such as seed potatoes, peas, beets carrots, spinach, chard starts, radishes, and lettuce. You can go online and google a North Texas planting guide. I find this helpful, but know that once your confidence is seasoned, you will find ways to bend the rules as you discover how to protect from freeze and create micro-environments where plants are happy.

Onion sets can be planted now without protection!

4. COMPOSTING

To begin with, you will need to know the basic role that carbon and nitrogen play in creating compost.

A good rule of thumb is to think of nitrogen sources as green matter – such as animal manure, food scraps, and fresh plant leaves – and then your carbon sources, known as brown organic matter – such as spent fallen leaves, brown branches, paper, cardboard, and even natural fiber clothing. These are the simple ingredients that, when heaped together, break down and make real soil! Another way to think of soil is as the forest floor.

Let's take a walk and see how this happens naturally! Follow me through the path of fallen leaves everywhere: most of these leaves have fallen from the deciduous trees and created a lovely layer of carbon upon the forest floor. So where does the nitrogen come from to break down these paper-like layers? Wait, I see a mule deer and her fawn trimming a huckleberry bush and leaving her deposit. Not only are the deer sequestering carbons with their manure, but the birds overhead are also liberally supplying their share. And under the leaves are fat earthworms weaving a trail of goodies. The worms not only create manure, but they also stir things up, adding more energy and life to the composting cycles in the forest. I don't need to mention all the other critters and plant matter that factors into the process; by now I am sure you have filled in the story. This simple illustration, though light, will help you understand the process of compost. But before we leave the forest, push aside a thick pile of limp, moist decaying leaves and scoop up a handful of earth! Mmm, can you smell the heady loam? – the very substance of what life emerges from! I can, and this makes me as much or more excited than my first flush of red ripe tomatoes grown in the compost I make.

Compost is soil that has started out as organic matter such as wood, leaves, paper, poop, food scraps, and natural fibers, that breaks down and at some point looks like what most people call dirt. This alone is the single most important task to achieve a great garden. If you start making yours in January, you still have plenty of time before spring arrives.

When compost is given care through turning, adding, and attention, it takes three to six months to complete the process. You can, however, use some of the partially composted matter before all of it has broken completely down. If you have a nose for compost you can smell the fertility when she's ready. If you do not trust your nose, go back to the forest and review how plants grow. Do you remember seeing that little moss garden atop the decaying log? Just keep going back to how life lives and thrives in nature when creating your gardens and you will have great success!

An urban composter is usually pricey and not necessary. If you are fortunate not to have hard clay or an abundance of rocks, start by digging a hole (but first check for lines, pipes, etc). If the ground is too hard to dig, then just start a pile using brown and green organic matter such as brown leaves, paper, (be sure to exclude glossy print due to petroleum in the ink), garden debris, and food scraps, excluding meat and dairy. Be on the lookout for bagged leaves that neighbors collect and for prepared-food establishments that will separate their leftovers. The larger the pile, the faster it breaks down because a larger pile naturally produces more heat. I know folks that have used their compost pile to heat water and their food. The best compost pile is at least 135°F. A better pile is 160°F. When getting to know your dirt, it's a good idea to get a compost thermometer so you can take her temperature!

Composting with worms can be done in all kinds of containers, including plastic tubs.

The ratio of carbon to nitrogen is 30 to 1 for making compost. That means it takes far less green matter, sometimes referred to as green manure, than carbon. This is good since carbon is more abundant. But equally living nitrogen sources such as grass clippings, table scraps, small animal manure, and plain ol' green weeds are abundant.

The heat from a pile with plenty of nitrogen ensures that any pathogens or weed seeds are killed. Piles that are slow to turn to soil may need more nitrogen or greens. If you need more green matter ask your local coffee houses to give you used grounds that are rich in nitrogen. In a nutshell, this is how you compost. How simple can you get?

Further discussion on the bad stuff: Unfortunately, we live in a toxic world and most of us who farm and garden are more aware of the poison factor than are our more disconnected communities. We also realize, though, that our food is our medicine and we make really brilliant

ANSWERS TO COMMON QUESTIONS...

My compost stinks! What can I do?

Add brown organic matter, grow mint around your pile, add microbes in the form of aerated compost tea, and/or sprinkle with diatomaceous earth, aka DE.

It looks ugly! I am afraid the neighbors will complain.

Cover with a tarp if neighbors complain, and make sure that you have your compost located close enough to the house that you use it, but out of site of onlookers. You may have to build or plant a screen around your pile.

What should I not put in my compost?

Inorganic materials that do not break down such as plastic, styrofoam, petroleum products, and metals. Also, do NOT add cheese, high fatty meats, or an abundance of citrus peels. A few peels do keep the vermin and strays animals away, though.

How should I add the compost to make a garden bed?

You don't need to till the existing soil; it's best to build. I learned this from a gardener down the street when I lived in Washington, whose garden was four feet off the ground just from composting. He gathered organic matter from the neighboring trees, canneries, barnyards, and God only knows where else. Though he was doing this lasagna style of gardening back in the early 80s, it wasn't until I moved to Texas and discovered ground as hard as cement and a shortage of earthworms, that I abandoned tilling altogether.

My point being: compost so that you can build your soil much like a forest floor is built. If you haven't the strength to dig a hole, no worries, just start a pile, or find a large round container with a lid to roll. Organic matter decays and turns into earth and the heat of the process will kill pathogens (the bad stuff — bacterium, viruses, or other microorganisms that can cause disease). Cover your pile with heavy plastic and for the Earth's sake, don't buy any — reuse plastic bags by fastening them together with duct tape to make a compost blanket. This will also keep rodents out and trap moisture and heat, which in turn speeds the process up. If you

choices daily. Worry is not healthy either, so do your best to understand that if you grow your own food free of pesticides and herbicides, it is already much healthier then anything organic you could buy in the grocery store. Fresh produce picked and then eaten is loaded with untold benefits. Besides the good you are doing for your body, family, and friends, you are helping our planet too. When making planting containers, use common sense. If you have access to raw wood, use that; if not, look for food grade materials and do not use creosote railroad ties. I use pallets and some may be treated, but I know the love will make up the difference! Enjoy the dirty life!

are doing this by yourself, consider doing a few piles so you can turn the heap since you may not own a tractor. (Did I mention gardening is great exercise?)

Common pesticide degradation is desired by those who only eat organic and they may rest assured since most or all pathogens are killed in the process.

Pesticide degradation is the process by which a pesticide is transformed into a benign substance that is environmentally compatible with the site to which it was applied.

Composting is well suited for pesticide degradation because the elevated or thermophilic temperatures achieved during composting permit faster biochemical reactions – thereby accelerating pesticide degradation – than is possible under ambient temperatures. The high temperatures can also make pesticides more bioavailable, increasing the chance of microbial degradation.

Some microorganisms may co-metabolize pesticides: the microbes rely on the feedstock for food and energy while breaking down an adjacent pesticide. Co-metabolism means that the microorganism does not receive any energy or potential food from the secondary reaction (in this case, from breaking down the pesticide).

Some creative composters utilized a shopping cart to screen large pieces from our compost pile.

The short end of this is that when gathering neighbors leaves, you do not have to worry whether or not they use organic practices. Once you have composted them, you and the soil are safe!

VERMICULTURE

Vermiculture is another way to compost. The basic difference is the worms speed up the process and you get to be a worm farmer. Unlike regular composting, you do not want to heat up your pile, but rather keep it temperate and moist for the worms.

There are a lot of high-tech methods for creating earthworm castings, better known as worm poop. Before I proceed, let me say that earthworm castings are agreed by most to be the best, most refined compost available.

There are different grades of compost, some being slightly more broken down. Take mulch for example. After you have spread it on your garden beds, a year later you will notice the mulch has begun to break down due to weather, water, worms and decay. Thus, you have compost, yet more than likely not all of the matter has decomposed and become compost.

This is a natural accruing event in nature and best observed as we did with the forest floor. But put that mulch, leaf litter, and food scraps in a sealed bin and stir for a good season, and you will find organic matter that is further along in the process — and for the purpose of veggie gardening, a suitable compost. The soil is more refined so the nutrients are readily available to your plants.

Finally, earthworms take organic matter, make a meal of it, and out comes refined compost. This is the bomb.

In addition to having refined compost in your gardens, bear in mind that raw matter, such as mulch or leaf litter, is still needed to keep the natural balance intact and create biodiversity in your garden.

Having said all that, you do not need a lot of space or money to have an earthworm bin, also known as vermiculture.

Here is my simple method for any eager soul. You will need:

+ A container (for inside or out) — most anything is acceptable, keeping in mind the little creatures need air. I use an old ceramic pot or a retired trash can, a rubbermaid bin, or even a five-gallon pickle jar.

+ Garden-variety soil, a few handfuls

+ Scoop of leaves

+ Paper scraps or cardboard

+ Worms — red wigglers are up to $30/lb but since you will most likely start small, just go to an organic garden shoppe and buy a can full, or search out in your site. You need the red wiggler worms; not just any garden variety hermaphroditic will do. This worm is not naturally found in soils since it is a species that has adapted to rot and decay. Originally a European native, now this wonder worker lives all over the world except in Antarctica.

+ Kitchen scraps that will make your worms smile

+ Acceptable scraps: veggies, fruit, paper, coffee grounds, eggshells, bread, pasta, rice, legumes and natural fibers

+ Not acceptable scraps: meat, cheese, fats, plastic, Styrofoam, and metal

You may have heard the expression, "eat poop and die." Well, I think that comes from a worm farmer.

As long as you feed your worms, they will poop. If not fed, they will eat their poop and wither away.

It takes ninety days to get a good harvest of castings in a small container, more or less depending on how many worms you start with. Keep an eye on them, get to know your new pets, and don't be afraid to see what's going on in their world; this is how you will learn. To gather up the castings, give your worms something they really love, like a rotten leftover you forgot in the fridge. They will swarm to this delicacy like flies to meat. Once they are centralized, you can take them out and harvest your castings.

Castings are the crème of the crop of soils and should be used sparingly. I use them like cupcake sprinkles and dust my entire site with a nice thin coating, especially during seed planting, since castings by nature are microbial-rich. When plants are suffering, a handful of castings does wonders too. You can imagine how helpful it is to have a couple compost sources going at once.

Finally, perhaps you have practiced composting, but have not yet had success. Most communities have master composting programs, or find a community garden to volunteer and learn with seasoned dirt makers!

Vermicomposting with kids can be great fun!

Journal Space

February

Mr. Fukuoka believes that natural farming proceeds from the spirit health of the individual. He considers the healing of the land and the purification of the human spirit to be one process.

—Masanobu Fukuoka, *The One-Straw Revolution*

OVERVIEW

February is the trickiest month of the year, a tease, and a torment; try your hand in a dice throw of natural gardening in February. This month is a great time to start cool season plants and seeds, but only if you have a plan of protection. The highs and lows of this month remind me of a menopausal woman, so handle February with care.

1. DECIDING WHETHER TO PLANT OR NOT TO PLANT

FEBRUARY CHECKLIST

☐ Decide whether to plant or not

☐ Prep the ground

☐ Protect your existing crops and livestock

☐ Start planting

If you don't like the weather, just wait a tad. Before the hour passes, you will more than likely get the change you wanted, for the moment anyway.

After living in North Texas for nearly 27 years, I am still amazed by the extreme weather changes we get in February. It still takes me by surprise when I am planting peas on a mild day in the sixties, then out of nowhere within minutes, the northern breath can drive me to wool clothing and a fire while the temperature drops into the twenties. If this weather does this to me, the least I can do for my plants is to give them cover and a good drink (of water) to lessen the shock.

2. GROUND PREP FOR COOL-WEATHER VEGGIES

You can take advantage of geothermal effect when planting your garden if you are blessed with pliable ground. We rarely have freezing weather that lasts long enough to freeze two inches beneath the ground, so an eight inch depth below surface, in essentially a sunken bed, is a good place to protect early starts. I'm not saying to cover your plants with eight inches of dirt, but rather to plant in more of a ditch than a raised bed.

Raised beds are great for other areas of the globe, but in truth are not always the best for crops in such a diverse climate as ours. If all you have is the capacity for raised beds, there are still plenty of crops you can enjoy year-round. Over-all, raised beds and container gardens do best here in the early fall and spring months.

Dig your furrow deep, about the length of your hand, six to eight inches. Down in the furrow, plant your seeds and then cover with the appropriate amount of soil for your particular seeds, found on the package. For example, lettuces will be covered by only a dusting of soil, whereas beets will be covered by an inch. But they will still be growing below surface level. As the seeds sprout, move along, adding more compost.

These hoop houses with plastic protect plants in winter and also double as a chicken tractor or chicken barrier.

Understandably, you may not have this option as a first-year gardener with possibly very hard ground and you will have to plant above ground anyway. In that case, protect your plants with straw or hay that is moistened so it won't blow away and/or have cold frames handy when the weather dips with extreme cold temperatures. If it's in your budget, check at high-end garden stores or some online seed companies (try johnnyseeds.com) for frost-proof cloths that are specifically designed to protect against frost.

For those with a smaller budget, dead leaves also work great, or even use old wool clothes. Save your moth-eaten sweaters for cover cloths! Then when the weather warms up a bit, you can remove the protection. Bear in mind though that when your plants emerge, they will be tender at first. It takes a few weeks to harden off (toughen up) to the weather.

In the event that you don't have pliable soil for a sunken garden, then you should consider a lasagna bed, or hugel bed, that is built above the ground. I'll talk a bit more about that in June.

As your early crops subside, fill the furrow with a mix of compost and earthworm castings so subsequent seeds will be well nourished. If you don't have castings, don't fret. Compost makes a great fertilizer. Since the weather is still relatively cool, grass clipping, bunny manure, and coffee grounds can be added.

Onions are tough and easy to grow. No furrow is needed. I like to make a small mound, though. Use short day varieties to start the bulb process when the daylight reaches 10 to 12 hours. Obviously, the earlier planted, the larger they get.

To protect your plants from harsh weather, here are a few simple, low-costs tricks:

✦ Water – the moisture adds a protective covering and is a must on new growth and all your plants to avoid freezer burn

✦ Fallen leaves – but you need to secure them with wire or blankets; otherwise the wind will blow your insulation away.

✦ Straw holds in place very well when stacked heavily

✦ Mulch, pine needles, cedar, hardwood, etc.

✦ Old clothing made of natural fibers

✦ Plastic vapor barrier and large landscape staples to secure to the ground

I keep a vapor barrier of garden frost cloth on hand for young tender growth, which I first cover with organic matter or straw, if handy. Then I cover the organic matter with the barrier and use large rocks to keep it secured. If you planned ahead and made a cold hoop frame, you are covered. But even if you are covering every stitch of plants, still make sure you soak with H_2O.

There are several examples in trial gardens as well as online of cold frames. Pinterest has lovely arrangements. Plus, some of our pictures taken at the farm on 8th Avenue can be found on the Elizabeth Anna's business page.

3. WHAT TO PLANT, AND WHEN?

Remember there are other things you can plant, but this book is based on easy successful crops in north Texas.

JANUARY TO MARCH 1ST

Onion sets – you will find these at your local nursery or feed store in bundles. They look like aged green onions that have been cut in half.

February is a hard month to plant, so you may opt to wait until extreme weather passes. This will give you time to chart out what you are going to grow for spring. Your chart can go in this book by just using a ruler; please do not think you need to be a pro on excel. Simply draw a diagram of your garden spaces and areas of guild and write in what you want to plant and the date. Unless you plan on preserving your fresh produce or giving a lot away, it is wise to plant seeds and starts in succession. I recommend you plant the same of similar crops about two weeks apart. This way you will get to enjoy fresh produce gradually and not all at once. Carrots I like to do longer stretches in time since in the early spring they keep well in the ground if not disturbed.

A note on cruciferous veggies: If we are having a warm spring, avoid over planting as aphids come out then and love to land on broccoli, cauliflowers and cabbage, making it a meal before spreading to their neighbors. When I get an infestation I usually harvest, rinse well, and eat these plants while they are young. For this reason I prefer to grow cruciferous plants in the late fall and early winter. The best bug control is diversity, so mix your cool season plants up with onions, marigolds, and other herbs for better pest control.

- Peas
- Carrots
- Radish
- Beets
- Parsnips
- Collards
- Spinach
- Chard
- Mustard
- Assorted Greens
- New potatoes – red or white (in the event of late freeze, be ready to cover new growth)
- Onion sets (seeds are not for the beginners)

- Cabbage
- Brussels sprouts
- Kohlrabi
- Kale
- Misc. black berries
- Grape vines
- Sweet peas for their beautiful scent
- Spring garlic – white or silver skins

For the daring!

- Lettuces
- Bok Choy
- Cauliflower
- Broccoli

Garnets in the Garden

As a young girl, I vividly remember gathering early crops with my grandmother for the evening meal. She would let me search for the young carrots and new potatoes.

With anticipation of the cream sauce these tender morsels would bask in, I dug into the ground uninhibited with my bare hands. My grandmother was a firm believer in "a little dirt won't hurt," and her hunches were true, as I prefer root veggies to the others. And for good reason, as they offer super nutrition and store longer and better than their above ground friends.

Then as a young woman in my twenties, I extended this love of root vegetables to include beets. I grew up hating beets since all I had tasted were pickled, so I just grew them for their greens, innately knowing that red stems were a source of life.

"Well," I reasoned as the beets got away from me and looked like huge garnets in my garden, "Lets eat one." I wrapped my mouth around my first fresh steamed beet with a bit of vinegar and butter, and the rest is history. Now I twist arms to get folks to try fresh, steamed, roasted, or sautéed beets.

These days, a frequently requested treat at our table from friends and family are my signature beets.

Do yourself and your gardening confidence a favor and grow onions! Without a doubt one of the easiest crops to grow, they do not need to be covered. They take wind, ice, freeze, and pop right back up. Besides that they help repel bugs, onions can be left in the ground for months. In fact, it is always so fun to find those onions I planted in February that hide and show up in November when the garden's large vines and leaves have withered from cold.

4. A NOTE FOR THE FARMER-GARDENER THAT HAS LIVESTOCK

Raising chickens and goats has become the new gardening trend, hence, the birth of the urban farm. Livestock is covered more thoroughly later in the book, but is mentioned here in February because perhaps more important than protecting your crops is protecting your animals. Most small farm animals are hardy, but if it gets below 25°F, have weather-barrier fabric, a shed, or large bushes for them to take cover. Young animals that have not flushed out their feathers or fur will need a heat lamp and a heated room if they are just babes. Goats are very hardy to cold, but they absolutely need a wind and rain barrier, such as a small shed or large doggy igloo with straw. Just as you would protect your crops with mulch and fabric during a cold spell, your animals need to be treated in much the same way.

5. ROOT VEGGIES: A CURE TO ANTI-ROOTS EATERS!

BEETS

Nutrition

1 beet has only 35 calories and is high in folic acid (especially important for pregnant women).

Beets and beet greens also contain potassium, calcium and betacyanin (an antioxidant). The leafy greens are also edible and even more nutritious!

Nutrition Information cited from Sima Ash Wellness Center: https://www.healing4soul.com/Simple-Home-Cures/natural-properties-and-curative-benefits-of-beet-juice.html

Elizabeth holding delicious golden beets

Food Value

This vegetable is a good tonic food for health. It contains carbohydrates, mainly in the form of sugar, and it has a little protein and fat. Beets are taken in a variety of ways. The leaves, like all green vegetables, should be cooked with a small amount of water and for only a short time. The fresher the beets, the better the flavor and the quicker they cook.

Beet juice is considered to be one of the best vegetable juices. It is a rich source of natural sugar. It contains sodium, potassium, phosphorus, calcium, sulphur, chlorine, iodine, iron, copper, and vitamins B1, B2, C, and P. This juice is rich in easily digestible carbohydrates, but the calorie content is low.

Natural Benefits and Curative Properties

Beets are of great therapeutic value. They have properties to clean the kidneys and the gall bladder. Rich in alkaline elements, potassium, calcium, magnesium and iron, they are useful in combating acidosis and assist in the natural processes of elimination.

Anemia

Due to its high iron content, red beet juice helps to regenerate and reactivate red blood cells, supplies fresh oxygen to the body, and helps with normal function of vesicular breathing, i.e. the normal breath sound. It is extremely useful in the treatment of anemia.

The juice of the red beet strengthens the body's powers of resistance and has proven to be an excellent remedy for anemia where other blood-forming remedies have failed, especially for children and teenagers.

Circulatory Disorders

Beet juice is an excellent solvent for inorganic calcium deposits. It is, therefore, valuable in the treatment of hypertension, arteriosclerosis, heart trouble, and varicose veins.

Kidney and Gall Bladder Disorders

The beet juice, in combination with the juice of carrot and cucumber, is one of the finest cleansing materials for kidneys and gall bladder.

There is more, but I will spare you in the hope you will grow beets.

AMAZING SUNCHOKES, i.e. JERUSALEM ARTICHOKES

These heat-loving, drought-tolerant plants are a must for the North Texas gardener. I must admit I had never even tried these plants until I grew them. Interestingly enough, the Jerusalem artichoke (Helianthus tuberous), also called sunroot, sunchoke, earth apple or topinambour, is a species of sunflower native to eastern North America, and found from eastern Canada and Maine west to North Dakota, and south to northern Florida and Texas. It is also cultivated widely across the temperate zone for its tuber, which is used as a root vegetable.

Above definition cited from Wikipedia: https://en.wikipedia.org/wiki/Jerusalem_artichoke

These sunchokes were harvested in the fall.

Though this plant is clearly native, it is not cultivated widely and is hard to find. It took me growing them to find that once they have been harvested, they need to be eaten soon. Putting them in the refrigerator doesn't help. After lots of trial and error, I have found that these guys store best in the ground. What you do not eat one season will come back the next and the next after that. This is the good news and, for some, perhaps the bad. So make sure when you plant these tubers you are committed. Here in the city we have started planting these along alleys and greenways.

In addition, chokes are great forage plants for goats and, well, my horse likes them too. The birds also like the seeds that fall out of the flowers in late autumn.

When the leaves dry out, the thick, fibrous plant makes for a good fire starter. Likewise, the stems are wonderful for kindling too, and boys that love sticks will get more then they can handle.

These plants, similar to their relative the sunflower, make a great sun block to nurse other plants and animals, and then people can eat the roots! Brilliant!

Ordering them is tricky and, due to access, usually quite expensive. I found them to begin with in Oregon and now I may be the only North Texas supplier. The plant can be planted in February and really anytime of the year. This is why I call this an amazing plant. The great part is you only need a little due to the wonder of Helianthus tuberoses!

Pick a nice sunny spot and bear in mind this sun lover will get taller than you and take up a 2 to 3 foot area within the first year.

A tip on planting: just do it! If you want these plants to proliferate in February, install them deep enough so freezing weather does not affect them.

POTATOES

February 6th is said to be the magic day for planting potatoes, but in my first year to grow potatoes in the south, I got them in the ground in late march and had a successful crop. How? Lots of compost and love!

Potatoes are easy. The hard part is finding seed potatoes for a reasonable price without the use of genetic modification.

If you love potatoes, dig a small root cellar. Grow plenty and each year set aside enough for the following year.

It is better to eat rice than genetically modified potatoes, so do your homework and insist on naturally grown seed potato.

How to grow potatoes

In a water glass, a tomato cage, a raised row, a furrow row, an old tire filled with poor soil, a compost pile by accident, a whisky barrel — each time I was successful. I have even planted new potatoes a month late in north Texas for both spring and fall harvest. I find it easier to grow here in the south as opposed to the gardens of my youth in the northwest. The lack

of rainfall is a blessing when it comes to growing potatoes. Too much water results in rot, black spot, and an ugly potato.

My secret is to make a furrow at least 8 inches deep with plenty of leaf mulch to make up for the heat and humidity that will come before the potatoes are ready to harvest.

How to plant potatoes

It's simple: cut into thirds and bury. It's best if the sprout is up, but really it doesn't matter. Mr. Potato will get it right!

6. CRUCIFEROUS VEGETABLES

When we think of cruciferous plants, the cabbage family come to mind. They're not as easy to grow as our root veggies, but with ease, you can have the benefits of this super cancer-fighting food in bite size pieces.

Included are rutabagas, kohlrabi, turnips, and to my surprise, the easiest to grow veggie, the radish. Learn to like this super snack; it could mean your life. While not considered a true root veggie, arugula is also an easy to grow cruciferous cool weather crop.

Cabbage is one of the oldest veggies and continues to be a staple for many cultures, including ours! Though there are hundreds of varieties, the green and purple are the easiest to find, but now Napa cabbage has become popular and is a tried and true variety to grow. This crop needs some protection in freezing weather, while crops such as broccoli can handles a frost well and will put out enough heads to feed a village. Cabbage and their relatives are all low in calories and high in micronutrients, especially manganese, Vitamin B6, C, K, calcium, tryptophan, potassium and several more.

I have heard it said that when you eat cruciferous veggies raw, something about the chewing action releases plant chemicals that seek and destroy cancer cells in our bodies. I overcame my first cancer and chomped on these gems cancer-free for fifteen years! I am a committed believer in the connection between nutrition and health and, for good reason, I'm continuing these practices as I battle cancer for the second time.

A CONSIDERATION WHEN PLANTING CRUCIFEROUS VEGETABLES

You may read in local agriculture resources that you should plant cabbage-family plants in

February, but I actually prefer to plant them in the late fall, due to their susceptibility to warm-season bugs, especially whiteflies and aphids. You can plant this crop all winter long, up until the end of January, as long as you protect it through extreme cold spells. Once common cabbage and kale are established, they can withstand northwestern winds in the teens. So I utilize this exposure in spots where other plants will not survive in the winter at my city farm.

Valentines Project: Hearty Food to get through the Winter Blues

Items needed:

✦ Enough used wine bottles to create a heart

✦ Cardboard

✦ Box cutter

✦ Sand, about 5 bags

✦ Compost, 2 cubic feet

✦ Red leaf lettuce seeds

✦ Compost tea or seaweed mix

This can be made on your sidewalk or driveway for best quick results due to concrete's heat factor. Yes, this can be placed in your yard where plenty of sun shines.

Directions:

1. Cut the cardboard in the shape of a heart.
2. Add sand to 5 inches deep.
3. Push the bottles upside down into the sand to outline the heart shape.
4. Fill the heart with compost.
5. Sprinkle seeds over entire surface.
6. Water gently with tea or seaweed mix.
7. Keep moist till seeds sprout and take root.
8. You will see red sprouts within a week. If a cold snap comes, below freezing, cover with plastic.
9. After 30 days, you will have lots of greens.
10. Eat your heart out! Pick the leaves instead of pulling the plants out of the earth so you will get months of green snacks!

Journal Space

March

For, lo, the winter is past, the rain is over and gone; the flowers appear on the earth; the time of the singing of birds is come, and the voice of the turtle is heard in our land; the fig tree putteth forth her green figs, and the vines with the tender grape give a good smell. Arise, my love, my fair one, and come away.

—Song of Solomon

OVERVIEW

Beware of spring fever and resist the overwhelming compulsion to buy twenty flats of starts and blooming color spots, not to mention three pairs of brightly colored garden gloves and shiny new tools. But rather, take stock of your planting spaces that by now should have cover crops that can be turned under or pulled as feed for your small farm animals.

This is also the month to check your fruit tree blooms, determine which spaces are ready to plant, buy seeds (rather than starts) since you have entered prime planting time, and perhaps start a garden that is grown on the edge of your property or just near a driveway.

MARCH CHECKLIST

☐ Plant peas as an early spring cover crop

☐ Take preventative pest control measures

☐ Start tomatoes indoors or in a greenhouse

☐ Choose your crops

☐ Try planting by seed!

1. SEEDS OR STARTS? PLUS, TIPS FOR GROWING TOMATOES IN NORTH TEXAS

Before I begin, let me tell you a story. I, like most gardeners, aspire to grow lovely prize-winning tomatoes. Every year I would buy beautiful 4" tomatoes starts of early girl and beef master, and dream of warm, juicy fruit that tastes like the sun running down my chin, right off the vine.

But after my second year of failure in Fort Worth, I decided not to repeat the same mistake yet a third time. To make matters more tantalizing, my mom bought me a book on heirloom tomatoes, brilliantly filled with pages and pages of perfect specimens to make any gardener green with envy. The book got me stirred up, but I knew those perfect pictures were not enough to ensure perfection. So I further studied their origins and favorite conditions and, to my surprise, North Texas was not one of Mrs. Tomato's stomping grounds. I knew good tomatoes came out of the south, so I was not without hope. Tomatoes are tropical by origin and love the mycelium-laced acidic soils of the rainforest, as well as moderate temperatures. When it gets above 80°F, Mrs. Tomato holds back her fruit and grips on to her bloomers. "Ah-hah," I said, "I need to start my plants early so high temperatures don't come and stunt my girls, yet I need to protect them from any thing below 70°. Sounds like a northern California summer – oh well, we Texans can trick 'em!" I figured out to plant seeds in mid-December to get a bug plant by April. I sought out a small farmer who specialized in growing rare warm-season tomatoes, then ordered these tiny gems for a buck each. I kept telling everyone I was a tomato farmer.

Like gold, they came in a small paper envelope. Tucked away, I counted all 20 seeds for a dollar each. It was a sunny day in late December when I began the process. I first soaked the seeds in our aerated compost tea overnight. While that business was going on, I took the best-looking worm castings my husband had made for me, his wife, the self-proclaimed tomato farmer, and filled a flat with the best soil we had. Each seed was placed with love, barely beneath the surface in a 4" pot of soil and then kept on the warm concrete until the sun set. I warmed the inside shoppe area to 70° since I had read that tomatoes do not like to drop below the 70s, which can cause an early blight on the plant. Day after day, I did my practice with the tomatoes. When the weather started to cool further down, I sealed my greenhouse and moved a small cooker in there to keep everyone warm. By now, the seeds all had names and I could tell each one apart. As time went on, my girls

out-grew their pots and got re-potted into larger containers a few times over with earthworm castings and homemade compost. The nursery had Not For Sale signs on these brandywine and Cherokee purple lovelies. Folks were not happy about deprivation, but I was not about to give up on my priceless trial growing run. The long-awaited April day came after much temptation to plant these along with other garden varieties I put out from other growers — but true to form and science, I delayed the introduction to their final growing habitat until an early morning in mid April! These ladies had a place of honor in our small urban farm with ample space, and the site was prepped ahead of time with garlic to rid the soils of any fungus or blight. There were also lemon drop marigold seeds dappled about the plot to aid in pest control! In July, the long-awaited day came and with sheer delight, I pulled off my better-than-chocolate tomatoes to show the world!

In most cases, I use seeds over starts, for the simple reason that a start is a transplant and transplants have a certain amount of stress to work through. The other obvious reason is that you get a lot more for your money if you apply some patience.

The only starts I do use are tomatoes. Due to our region, we get late freezes and frosts nearly every year up to the middle and, recently, end of April, butted up next to high temperatures. Since tropical plants, such as tomatoes, easily get blight from dropping weather, I start growing them from seed in my greenhouse and set them out after the 16th of April. This can even be too early, as one year I waited and waited, and then during the last week of April, I did not listen to my hunches — and sure enough, we got freezing winds and my prized tomatoes that had grown to three feet tall and wide were stunted and eventually blighted. We ate the first flush of lovely tomatoes, but soon after, the signs of blight showed up in full force. I finally got tired of peeling off dead leaves and brown spots, so I yanked my ladies out of the ground and gave the plot over to the chickens until I was ready for a later fall planting of greens, which put the soil back in good health.

2. WHAT TO PLANT?

March 21st is the spring equinox, or the first day of spring.

This is the longest day of the year, so make use of all the daylight you will be blessed with. A planting party is in order!

For early March, see the February chapter's fuller list of plants, but for planting in mid to late March, use the following list.

This is an opportunity to gain confidence in your ability to grow seed. If you have followed the previous months' instructions on growing soils and creating living beds, then no need to worry. If you are still struggling with making good earth, no need to fear — just find a good compost source and/or use earthworm castings to frost the top of your planting area. Buy fresh seeds, which are much more likely to germinate than older seeds. To use up older seeds, it's fun to make seed balls using found clay from our North Texas ground and tossing the balls into a field!

HOW TO CHOOSE YOUR CROPS: AN OVERVIEW FOR THE YEAR

With so many choices, it is hard to make a decision about what to grow when you have limited space. I have found that we have more success with the things we are passionate about. So grow what you love, as well as easy crops, as these too will become what you enjoy. I have included here an overview of very easy crops to keep in mind, but not necessarily to be planted in March.

When deciding when to plant, don't forget that air temperature is not the only factor to consider, as the warmth or coolness of the soil also plays a huge part in growing success. The following is a list of both cool- and warm-season plants and crops that will give you satisfaction due to the ease of growing them. This list will expand for you as you grow in confidence with edible farming, but in the meantime, this is a good place to start.

The following list are edibles that are easiest to grow for most, along with their companions. Make sure to plant these during the appropriate months, as cited in this manual.

SEEDS

- Peas
- Arugula
- Radishes
- Green beans
- Heirloom melons
- Cucumbers

- Okra (once ground is 60°F)
- Basil (once ground is 60°F)
- Beets (before ground is 60°F)
- Onion sets
- Potatoes
- Lambsquarter

PLANTS

- Peppers
- Cherry tomatoes

- Blackberries
- Blackfoot Daisy

ALSO TRY:

- Prickly Pear
- Peach trees

- Passion fruit

Heirloom melons, peppers, cucumbers (try the Japanese variety), and basil can be started indoors now and planted in the ground once the danger of frost is past, or you can wait and plant the seeds directly outdoors once the temperature is safely above 60°F. Okra doesn't usually transplant well, so don't bother with starting it indoors.

Blackberries can be planted winter through spring, so make sure and don't wait too much longer to get them in the ground. The same goes for beets, which can be planted from fall until the ground hits 60°F.

Onion sets can be planted now, or later in the winter months.

Cherry tomatoes, originally found in the tropics as a wild vine, are a warm-season plant; they tend to naturalize once established in the garden, especially if you let your chickens range in them once the crop is finished. The chickens will not only clean up the bugs, but also will plant your tomato seeds for next season – somehow, mysteriously, those digested seeds keep dormant for 6 months, which happens to be the next warm season, and you will have cherry tomatoes everywhere!

Wild spinach, also known as Lambsquarter, is best to broadcast by seed in the fall for a spring and summer harvest, and once this wonderful plant takes root, you will have more than enough for yourself and your flocks and herds, as it naturalizes in the edible landscape without any help from human hands!

Texas native Prickly Pear is another great selection for the edible landscape and the unconventional garden. Its bright purple bulb-like growths, known as the "tunas," are tasty in Mexican cuisine, particularly salsas.

Blackfoot Daisy, another Texas native, is a must for the edible garden for its drought-loving charm and its vanilla scent, which is alluring to both bees and people.

Make sure and choose a peach tree with low cooling-day requirements. This is determined by how many days are below freezing during our winter and early spring months. Read more about transplanting fruit trees in the September and October chapters.

Passion fruit is a lovely hot-season thriving vine and a great pollinator and butterfly attractor. I like to use our native species, called Maypop, so that she comes back year after year.

A NOTE TO SEED GROWERS

Weed seeds will more than likely grow along with your desired seeds unless you have primo compost. But don't worry: let it all grow until your plants are highly recognizable, then wet the bed well and pick the weeds off. Toss the little green imposters on top of the beds as tiny green mulch and good fertilizer. They will not re-root unless your weeds are Bermuda grass; in that case, feed them to the chickens or allow them to scorch on the concrete or gravel.

MID- TO LATE-MARCH SEEDS

- Green beans
- Carrots
- Fennel
- Beets
- Greens, such as spinach, chard, kale, lettuce, and arugula
- Marigolds, both seeds and plants, so that the roots of the marigold can start their science and rid the earth of non-beneficial nematodes
- Dill — dill likes cool weather, so don't wait until April if you like to pickle your crops.
- Zinnias — it is too early for transplants, unless you want to cover them when the northwestern weather comes to wreak havoc.
- Basil, from now until late April

MID- TO LATE-MARCH STARTS

Feel free to try your hand at seed starting these herbs for a challenge, but it will take some patience on your part. Parsley is the least difficult to start from seed. Hardy herbs can be planted anytime in March, while annual herbs such as basil, cilantro, and patchouli are best started by seed at the end of March.

> **Hardy herbs generally are the woody varietes, such as rosemary and thyme. Annual herbs, such as basil, dill, and fennel, have softer stems from the start and often stay soft. But even these annuals are good reseeders, in that they reseed themselves in good soil conditions.**

- Rosemary
- Parsley
- Thyme
- Marjoram
- Oregano
- French hollyhocks
- Lavender
- Salad burnet
- Tansy
- Curry plant: loved by the English and used in their sandwiches at tea time
- Yarrow

It is not too late to plant container fruit trees and container berry vines!

Although, I do not recommend strawberries for the novice, as they prefer cold, moist, and acidic soils, and so are not a good North Texas plant. If you love how strawberries taste, try your hand at Kiwi Vine. They are native to China and are a much better choice.

Plant all mint in contained areas, as its extensive root system will take over. Do not plant basil starts outside until April 16th unless you plan on bringing them in when the weather drops below the fifties. Though cool temperatures may not kill your basil, it will stress her out and cause many of the leaves to drop.

MID-MARCH TO MID-APRIL SEEDS

+ Cucumbers
+ Eggplant
+ Melons
+ Pie pumpkins
+ Green beans (only by seed)

+ Summer squash (yellow crooked-neck, zucchini, etc)
+ Radishes
+ Cow peas
+ Heat intolerant lettuces

A lot of gardeners will tell you that tomato starts can go out after March 16th, but this does not account for the late northwestern winds we get up to mid-April, so I recommend you wait to plant tomato starts until mid-April to protect them from winter's last hard blow. The exception would be if you have a Southern exposure with good protection.

RECOMMENDED CROPS TO AVOID

Most anything in the cabbage family is prone to bugs once it warms up; plants in this family attract whiteflies like a magnet, which then spreads like wildfire to your other crops. I suggest you harvest your broccoli, kale, and cabbage for the betterment of your plant community. But, again, observe: you may be one of the lucky ones with great circulation, diversity, and the right balance to keep your garden free of these pests. So if your garden is looking good, count your blessings and enjoy bigger cabbage heads.

3. SPACE GUIDELINES

When it comes to space, think outside of the garden box. Look around and see if you have walls that face the morning sun, or cyclone fences, or even a crack in the concrete where cool season veggies will fair well. Or is there a tree that is bare in the early months where wild food can sprawl, and then enjoy the shade of the canopy later on in the summer?

Trees pull up and hold moisture in the ground, plus they do the work of mulching with their leaves.

A Time To Be Born

On a late March day in coastal Washington, when the temperature was in the high 70s, I sensed it was time to give birth to my second son, Ethan. A walk, I surmised, was just the thing we needed to get our now past-due baby a nudge to join the universe.

My husband came along to literally support a good thing – and me – because on the way back, contractions started hard and fast. Before the pain turned me back home, I spotted old-fashioned quince bushes full of red blooms, still in their early-stage mass; tight buds appeared stuck to the thorns of this lovely shrub. But my favorite early-spring show was around the block: with big bright bushes, a true vintage shrub known to bloom along slender weeping branches, the stunning forsythia greeted us all in ablaze.

These well established plants in this old neighborhood got me off track and before I knew it, I was staring at deep blue grape hyacinth popping up out of a virgin meadow (my neighbor's front lawn). I ventured out to capture a tiny

Quince in the alley

wild bulb for a moment, until a hard contraction had me bent over in my husband's arms. "Okay," he said, "I think it is about time we get back." And in just a few short steps back to our home, set with bouquets and a burning fire, I settled in for what was the easiest birth of all three of my boys.

I can not say for sure if it was the spring walk that ushered in peace, but I have an inkling that the day's beauty did something out of the ordinary.

Ethan now is in his thirties and spends some of his free time growing flowers and vegetables, but he especially loves the flowers he cultivates.

DEPTH GUIDELINES

Most seed packets give instructions, but for further success, factor this in: The soil can work both as an insulator of warmth and of cooling. So in the cold months of gardening, plant a little deeper than you would in warmer months.

As the days get longer and warmer and the weather is optimal, plant according to seed instructions. Then as the heat comes on, start the mulching process to keep the roots cooler and reduce the amount of water needed. All types of mulch will work. Here in Fort Worth I use Texas Cedar and a native leaf mulch that I make. I prefer pine mulch around my tomatoes, since our soils are alkaline and pine adds acidity. This is a good reason to grow a pine tree. If you don't have the room, ask your neighbors during the holidays if you can have their leftover Christmas tree.

Remember to keep gathering leaves for mulch or for your compost. Drive around your community periodically on trash pickup day and gather lawn bags set out!

VINES

The following is a list of vine plants that fair better with support, though allowing them to sprawl on the ground is fine if you have space and good drainage to keep them from rotting.

- Melons, all sorts
- Cucumbers
- Green beans
- Grapes
- Tomatoes
- Spaghetti squash
- Winter squash

To grow these without taking up a lot of space, use tomato cages, bamboo, homemade trellises or a fence line so these wandering plants can be part of your edible site. These plants are content to grow upward, outward, and sideways. Garden supports prevent the common problem of root rot and keep crustaceans, such as rolly pollies, from eating the outer layer of your crop.

OTHER SPACE FACTORS, AND PEAS AS A COVER CROP

The space your plant needs is dependent on how fertile the soil is. Really take a hike in the lush wild and you will find a vast variety of ground covers, perennials, shrubs, trees, and annuals all living in harmony together, producing and thriving. Over-crowding is not a problem if your flora has a great base, but for new gardens, follow the recommended instructions and keep adding compost so that you can grow plenty.

Plant peas in the early spring to add nitrogen to the soil and then you will have a great prepped area to plant tomatoes later in the warm-season months. There's a lot of controversy over whether or not to pull the peas out or till them into the ground. They actually fix nitrogen either way, regardless of which you choose. It is advantageous to pull them and lay them back on the ground where they were to act as a green mulch and further aid the soil, or give them to your goats to eat. Compost them if you really don't like the messy look of green mulch, but I wouldn't bother with tilling them into the soil.

If you want to grow prized large tomatoes, then by all means give this queen her space, or else you will herald in all sorts of bugs and other problems.

But if you want to have a wild-style food forest, plant small sweet cherry varieties that have long vines, such as grape tomatoes, and you will have the entire flavor without all the work!

Otherwise, plant your fair-prize-sized tomatoes with plenty of space — anywhere from 3 to 5 feet, depending on the variety.

Okra gets really tall, so plant a tall variety by seed on the west side of your patch to protect your prized tomato plants from the harsh sun rays of Summer's fury.

I usually keep my best plants in the greenhouse until about the middle of April, and then find the perfect spot so that by July, tomatoes as big as softballs hang on the vine.

Food forests are fun to create and allow Mother Nature to do her thing.

4. EARLY PEST CONTROL

+ Take preventive pest control measures from the start.

+ The best antidote for bug control is healthy soil.

+ Plant marigolds to clean the earth of non-beneficial nematodes.

+ Buy beneficial nematodes to biologically control ants.

+ Don't make it easy for bugs to track down all your goods: mix it up with flowers, patterns, different heights, and a variety of plants.

+ Start out with healthy plants!

+ Do not buy plants from box stores, who are often made to throw out all their stock due to diseases.

+ If you do buy box store plants, wash them with aerated compost tea.

+ Do not use pesticides, since they kill everything – not just the bad bugs.

+ Use aerated compost tea to bring your site to life with healthy microbes. Simply put, compost tea feeds the earth and plants and then continues to regenerate. Kind of like when you eat raw yogurt, the good bacteria spreads. Likewise, when you nourish your site with aerated compost tea, it spreads and seeks out bad bacteria and pathogens to destroy, while enhancing soil fertility at the same time. Locals can buy this from our city farm. If you are not close to a source, look on YouTube for tutorials.

5. STARTING AN EDIBLE EDGE

This is my favorite theme and I hope to someday devote an entire pamphlet to edge gardening. But for the time being, if you have not already noticed some advantages to edges, here is a list of things to look for in a location:

1. A place of energy
2. A place of transition
3. A place nutrients gather
4. A place the wild birds congregate
5. A place that collects leaves, debris and surprises
6. Perhaps feature a fence to create a natural guild of plants
7. Provide definition and privacy.

An edible entryway edge garden

6. TRICKING YOUR CHILDREN TO EAT THEIR VEGGIES

I had never heard of a child who didn't like fruit until Isaiah, my youngest son, came along. I would even put sugar on his apples to get him to eat nature's treats. But to no avail, unless the fruit was well disguised, I was unsuccessful.

My granddaughter Melody on the urban farm.

Then we moved to our new home, where I installed a victory garden and planted veggies and melons. At the time, Isaiah was young and eager and helped me plant a nice mound of cantaloupe. Their emergence mesmerized him and kept him coming back to the garden daily for inspection. It was nearly impossible to keep him from picking his pet project, but somehow I managed, and the suspense grew, as did his anticipation to taste the fruit.

And so it was, at the age of four, that Isaiah found his first fruit.

Recently Isaiah, now in his twenties, humored me once again as he chopped into my heirloom honey melons and said to me, "Mom, this is the best melon I have ever tasted." Of course, like any corny mom, I reminded him of his first fruit.

Point proven: the things we grow, we come to like. And better yet, our picky kids will eat what concerned moms feel they need.

7. COMMON MISTAKES

The most common mistake first-time gardeners make is planting starts versus seeds. Seeds sown directly into the ground have the advantage of a good start without the stress of a transplant. The roots are never restricted and establishment is easier from the beginning.

If you are one of those people who say that seeds don't like you, change this to, "I am a brave gardener and my seeds will sprout." Go ahead and belt it out while you are gardening. To further your successes, soak these babies in seaweed or compost tea overnight. As the weather gets hotter, seed planting gets more difficult for obvious reasons. So for gardening success, plant seeds early and say nice things to your plants.

It is worth repeating: make sure the soil you are planting your seeds in is alive and has nitrogen. In other words, if your compost is more mulch, leaves, or straw than fine soil, you have a deficiency on your hands and this explains why your seeds do not pop.

Over-watering leeches plants of soil nutrients, causes fungus, slows down root development, and costs a lot of money, besides being wasteful. Give your plants water if they are wilting. The more rain water the better, since city water has chlorine. However, if you fill containers with city water and let sit for 24 hours, you are essentially distilling your water of most of the chlorine.

Gardening with gloves all the time is not as good for you or your plants. Multiple studies have shown the healthy connection between the earth's surface and our well being, not to mention the strong relation to our vital immune system and gut. I like to garden barefoot to even further my grounding!

That's right, I said it! Gardening bare handed is so good for you and the plants. You are both alive and need each others' touch. I find that novice gardeners do not touch their plants enough. This is one of the reasons plants do not thrive. In the wild they have the birds, bees, bugs, and deer, but in your site they have you. So make it a habit to give your plants warm fuzzes, and plant plenty of bee and butterfly attractors.

AFRAID TO MAKE MISTAKES?

Please just have fun — so what if some things fail? You may learn that your problems provide you with the solution.

Journal Space

Garden Plans

SPRING

April

There is no spot of ground, however arid, bare, or ugly, that cannot be tamed into such a state as may give an impression of beauty and delight.

—Gertrude Jekyll

OVERVIEW

This month is almost your last chance to install seeds and starts before the weather's heat makes starting plants a major challenge. By now, most of your seeds and starts are taking shape, and if you planted root crops such as potatoes, beets, and carrots, harvest may be at hand. At least pick some baby peas, carrots, and potatoes, and smother in cream sauce for an early spring meal!

1. WHAT TO PLANT?

By now the season is underway, so most will opt to plant starts. If you are lucky, some of the garden nurseries will have mature plants that you can pop in a container or bed, but bear in mind that it will take some babysitting for these older plants to take hold. In the meantime, I have had success planting the following by seed and start. Remember that if your earth is tired, your plants will prove weak and may get attacked by warm-season bugs. So by all means, before you add seeds or starts, amend your soil with at least an inch of quality compost, earthworm castings, or organic fertilizer. Later on I will discuss more about fertilization and edibles. For now, here are my best recommendations for success!

> ### APRIL CHECKLIST
>
> ☐ Plant more seeds and starts – don't forget okra!
>
> ☐ Start bringing tomato plants outdoors
>
> ☐ Check if your root crops are ready for harvest
>
> ☐ Thin out root beds if too crowded
>
> ☐ Weed and mulch your beds

- ✦ Eggplant (seeds or starts)
- ✦ Okra (seeds or starts)
- ✦ Basil (seeds or starts)
- ✦ Sugar baby melons (seeds anytime in mid to late April)
- ✦ Cantaloupe (seeds anytime or starts in mid to late April)
- ✦ Golden purslane (seeds)
- ✦ Black-eyed peas (seeds, not starts!)

- ✦ Lemon balm (starts)
- ✦ Bee balm (starts)
- ✦ Patchouli (starts)

2. GROWING TOMATOES

Everyone, it seems, wants to grow tomatoes. It's understandable, as there is nothing like a mouth-watering, zesty, homegrown tomato. But be warned: growing tomatoes is not for the faint at heart. Everyone should grow some for enjoyment, but until you have learned some of the rhythms of nature and concepts of natural gardening, don't put all your time and treasure into one crop.

Despite the occasional problem, growing tomatoes is totally worth it.

Tomatoes are high feeders, prefer a higher pH than most veggies, and tend to be on the finicky side. For you science-oriented folk, tomatoes like their soil pH around 6.0 to 6.8. Briefly, pH is a measure of soil acidity or alkalinity. On the pH scale, 7.0 is neutral; so the range that tomatoes prefer is slightly on the acid side. (By the way, that's the pH range at which most vegetables grow best.)

Basically, they need lots of great organic matter. I like to mix a variety of amendments in hope of satisfying my tomatoes.

Here is a simple recipe to feed your tomatoes:

- ✦ 60% garden-variety compost (made from leaves, greens, and mulch)
- ✦ 20% earthworm castings
- ✦ 20% equal parts of:
  - ✧ coffee grounds
  - ✧ green sand
  - ✧ lava sand
- ✦ Plus, seaweed in liquid form found in garden stores, or fresh compost tea, to water with afterwards

This recipe intentionally does not measure exactly – but you get the picture. This is not as much a science as an art. If you do not have time for all this, find a good organic label you can trust and amend with an alfalfa-based fertilizer.

SOME TOMATO TIPS

- ✦ Tomatoes need sun most of the day, so avoid growing them in a shady spot, despite what you may read or hear.

- ✦ Tomatoes are not drought tolerant, so plant in thick compost, add plenty of mulch, and add water amenders such as grow core, shale, or decaying matter to save on water. They need to stay moist, but not muddy. This is why a thick layer of compost, up to 10 inches, can be a safeguard for your plants roots.

- ✦ Tomatoes prefer rain water. Too much hose or irrigation water will make them sick, so gather rain water or put a filter on your outdoor spout. They also do not like their foliage sprayed, so water by hand or use a soaker hose.

- ✦ Tomatoes are tropical plants that prefer weather like is found in northern California, but with a little adaptation, the right plants, and positive energy, they can do well here. So don't put them out too early, as cool weather will stunt or kill them.

- ✦ Spring tomatoes do not handle frosty weather, but tomatoes that make it through to the fall – now, those are some tough girls, and they can handle a light frost due to a term called "hardening off." As the name suggests, they have hardened off to the weather conditions and are a stronger plant.

- ✦ Blight is a common disease that preys on weak tomatoes, and when it comes, it is here to stay! Tomatoes are prone to blight at under 60°, so avoid too much exposure to cold days in March and April.

- ✦ Start seeds inside in a warm, sunny area where the temperature does not drop below 60°, or buy mature plants and set them out in mid-April to be on the safe side. (As you learn intuitively, there will be times they can go out in March.)

- ✦ It is not a bad idea to keep your tomatoes in pots that can be taken out and brought back in until all chances of frost are gone.

- ✦ There are so many varieties of tomatoes that it will make your head spin, so for success, stick with heat-tolerant and heirloom varieties.

- ✦ Tomatoes don't like our July and August weather, so time your planting, or provide a shade cloth to cut out 20% percent of the hotter-than-hell UV rays. This will still provide them with plenty of sun. Or do what I do and plant sunflowers as a nurse plant in your tomato crop. Not only will your tomatoes enjoy shade, but in late summer, you will enjoy the red cardinal's song.

- ✦ For success, grow small tomatoes such as cherry, grape, or any patio tomato. Since these mature early, they are less likely to be plagued by common tomato pests and problems.

- ✦ My favorite tomato to grow in North Texas is brandywine. Number one, it is one of the most delicious crops available; and number two, this heirloom can take our hot summers. I have personally harvested double-grip over-sized tomatoes in late July, when it is hot as a habanero.

- ✦ If your tomatoes are not happy, they will let you know by turning yellow, getting bugs, rust, and just plain turning ugly. So give your tomatoes what they love. Or just grow them for the fun of it without all the fuss and see what happens. You may have beginner's blessing!

3. TOMATO PESTS

When it comes to fungi and pests, the best prevention is diversity and healthy soil. But you may still come up with some problems. A low-cost treatment for most fungi is to mix 1 tablespoon of baking soda per quart of water in a spray bottle and spritz it on the tomato leaves in the morning while it's still cool. Repeat the application several more times if the fungi remains and also sprinkle organic cornmeal around the roots of the plant.

Here are some common tomato diseases found in our area and more specific treatment suggestions.

ANTHRACNOSE is a very common fungus that causes tomato fruit to rot.

Symptoms: Small, round, sunken spots appear on the fruit. The spots will increase in size and darken in the center. Several spots may merge as they enlarge. The fungus is often splashed onto the fruit from the soil. It can also take hold on early blight spots or dying leaves. Wet weather encourages the development of Anthracnose. Overripe tomatoes that come in contact with wet soil are especially susceptible.

Management: Copper sprays offer some resistance. Remove the lower 12" of leaves, to avoid contact with the soil. Don't water the leaves, but only the base of the plant.

BACTERIAL SPECK

Symptoms: Tiny, raised, dark spots, usually with a white border.

Management: Apply copper fungicide at the first sign of symptoms, according to manufacturer's instructions.

BLOSSOM END ROT is generally attributed to a lack of calcium during fruit set. This could be caused by too much high nitrogen fertilizer or uneven watering, resulting in fluctuations of nutrient availability.

Symptoms: Dark brown/black spots develop at the blossom end of the fruit and enlarge as the fruit rots.

Management: Remove affected fruit and provide regular, deep watering.

EARLY TOMATO BLIGHT is another fungus that can be soil born and is found in transplants.

Symptoms: First signs are yellowing leaves and brown spots.

Management: Pick affected area and treat soil with microbes, such as fresh compost tea. Apply copper fungicide spray.

LATE BLIGHT is similar to the others mentioned, but occurs later on and is caused by too much moisture.

Management: Treat the same way as with early tomato blight, and to avoid it in the future, make sure your tomatoes have plenty of room between them and well- drained, healthy organic soil.

4. AMAZING OKRA!

Need satisfaction? Grow okra. Please learn to like this no-hassle plant, as it is without a doubt North Texas's most sustainable crop. It uses very little water, grows in the heat of the summer, provides food daily, requires little fertilization, and is tasty. Okay, I hear a lot of you complaining you don't like it. Well I never did either, and I loathed it until I gave it a try. This last year, Dan, a fellow garden geek, treated me to some heirloom seeds. They turned out to be jolly green giants. Thinking if I disguised the flavor I could stomach this southern favorite, I whipped up an herb-infused primavera dish with lots of pulverized okra. We loved it, as did customers at our Saturday market. Yet beware, because this girl can get away from you. I now find myself walking in the shade of her beautiful leaves and hollyhock-like-flowers, munching down on tender pods — raw! That's right, raw! I later found out this miracle food lowers cholesterol and is loaded in vitamins.

Besides all this, okra is a great support plant for tomatoes in the summer when planted to aid your juicy red fruit from the hot west summer sun. Plant okra on the west side of your tomatoes and they will tower over as the summer comes into its fullness. This holds true for other sun-sensitive crops, such as blueberries, green beans, cucumbers, pumpkins, and summer squashes. Plant okra seeds in March and then plan on placing tomato starts once you see the okra sprout. Other crops can be seeded at the same time.

Besides the wonderful food, okra produces pods that, if they get away from you, will hold tough, mature seeds that can be roasted like chicory and enjoyed like coffee.

5. HOW TO CHECK IF YOUR BURIED CROPS ARE READY, AND SOLUTIONS FOR OVERCROWDING

Often the problem is the solution, so while your carrots and potatoes and beets may be pushing each other right out of the ground, these early crops may just crown at the soil's surface and this is your only hint of what to thin from your crop. Before you do this, water down the area really good so the roots of the plants you don't want to pull out will stay in the ground. Then, after pulling the other veggies, spread compost back along your row or patch and re-water. This will ensure the plants you have not pulled will keep growing and you may just get some prize-sized roots!

A warning about root harvesting: Be careful not to dig into the ground, but rather use your hands to brush the soil aside, revealing where to pull. By doing this, you will avoid slicing your beautiful veggies. It is best to thin when the ground is moist, then again, let the other crops stay put about two weeks or more for larger produce. I try to thin evenly, but if the planting is not even, then thin the obvious.

Carrot harvesting with children is heaps of fun!

Don't worry about getting everyone out, as most root crops are quite content to stay down under. I love it when a carrot gets passed over and then later in the season, this beautiful flower emerges that most think is Queen Anne's lace. The beauty of leaving a few stragglers behind is that you will then have your own seeds to plant next season or fall, which is just around the corner. To collect the seeds for replanting, they must dry out on the plant completely. If you are going on vacation in late July, tie a paper bag on the plant's seed head so the seeds fall inside, or let them fend for themselves as volunteers. This practice is common in food forest gardening and takes a lot of observation to become a master.

6. WEEDING & MULCHING

It is natural to weed while you are thinning, unless Bermuda grass is so entangled in your plants that you dislodge their roots by pulling the runner. In this case, just make sure to add more compost due to the competition for nutrients. Later, when you harvest all your crops, you can make sure to get to the root of the Bermuda. The best solution for getting rid of Bermuda is to weed it on a regular basis. You can slow her down with agriculture vinegar, sheet mulching, and other methods but in the end, the best method I have found for managing this invasive grass is raising chickens. Think of it an an opportunity — now you have a natural feed source!

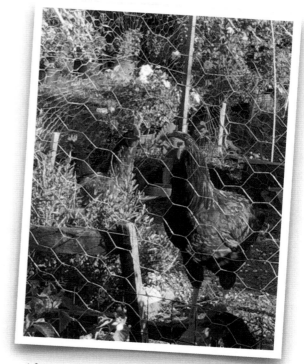

After the crops have been harvested, the chicken wire serves as a pen where these roosters are gleaning the leftovers and finding all sorts of worms and bugs!

Once you do remove weeds from growing spaces you will need to cover them with mulch, hay, or straw. Ground wants to be covered and if you do not assist her, Mother Nature will do it herself. Mulching with carbon-rich materials not only inhibits weed growth, but also helps with water retention. Win-win!

Some crops also act as mulch — living mulch! Sweet potato vines, black-eyed peas, and other cover crops keep other seeds from taking root and also help hold moisture in the ground.

Green mulching is a fairly new concept. You simply pull the large green mass from your plants and lay them on the ground's surface. For example, I grow comfrey for several reasons, and one of which is to use the large-biomass leaves as mulch in my food forest. This is an amazing plant that needs shade and moisture, so if you have that type of spot on your site, make sure and grow this multipurpose plant. It is a challenge to get her to take, but once you do, this herb will flourish! Her beauty comes from her relation to the borage family, so you get the benefit of gorgeous blue flowers, too.

Journal Space

May

I hunted curious flowers in rapture and muttered thoughts in their praise.

—John Clare

OVERVIEW

Mother's Day is aptly dated in the month of May, a time when life is bursting forth everywhere. My granddaughter was born in May; Melody May perfectly personifies this month, as both are free-spirited and bring joy to every corner of creation.

In the farm and garden, keep your eyes alert for new peeps. Just a few days ago, we chanced upon a small black hen that hid its nest among the thicket of ripening blackberries. To everyone's surprise, the mama now proudly parades the chicks near their hiding spot. In May, the goats kid, cardinals forage for their new young, and even forgotten planted seeds are now in bloom, above the fertile soil. A new song can be heard among our winged friends, for spring has opened wide. This month is a reminder to be on our toes, as a green peach can turn right into jam on the branch – from fruit to preserves in a breath. So be ready to make haste as Spring throws you her best!

> ### MAY CHECKLIST
> ☐ Take your shoes off and get in touch with the earth!
>
> ☐ Plant more seeds and starts
>
> ☐ Control your weeds
>
> ☐ Remember to water the garden!

1. THE INTUITIVE GARDENER

Intuitive gardening goes beyond textbook gardening and is instead built on the foundation of our inherited relationship to the earth and the earth's creator. As we learn to reconnect and feel our natural surroundings, this sensitivity is sure to come in handy: from when to plant, what to plant, and even how to live in peace with wild-life, non-beneficial insects, and our

A mother duck with her babies

neighbors. Though this is tricky to teach, I am sure a lot of experienced gardeners do this naturally.

As an intuitive person, I hope to expand your gifts and provoke your inner desire to frolic in the garden of nature.

What I am talking about is not a new-age idea; in fact, in our not-so-far-off history, George Washington Carver, famous for the discovery of peanut butter, practiced intuitive gardening as a way of life.

The following is an excerpt from: http://www.intuitivegardening.com/intuitive_gardening/carver.html

George Washington Carver (b. 1864 - d. 1943) was an American scientist and educator. In the early 1900s, monocultures of cotton crops were grown in the southern United States and it was depleting the soil… in the little book called "The Man Who Talks With the Flowers" by Glenn Clark, I found out that he derived the ideas for these products by divine inspiration. Carver would be up at 4:00am and go into the garden to pray and be with nature. He'd connect with the plants and meditate on them. Carver would then receive an answer from God, the plant, and Nature, and then would go into his laboratory and follow the instructions and inspiration to create these products.

Intuitive gardening, for me, is not only learning to be in nature and to receive inner guidance, but also connecting to our own tamed sights and sounds to find the right stuff to grow.

I know at times when I cannot get my head clear, just by digging in the ground and tending to my crops, the answer comes so swiftly and beautifully.

I believe life lessons from the garden are the best lessons because we learn to be physically grounded and free from all the negative energy waves of modern life. We can enhance this experience by gardening on our hands and knees or just by taking our shoes and socks off!!!

Let's be mindful of the way God gives us substance – through the earth. When we connect to her/his bounty, we are nurtured and literally fed.

As a confused youth I headed to the woods, where I took handfuls of moss and breathed them in until I felt at peace. Recently my mom sent me a candle called "forest floor" – the outdoors give comfort not found elsewhere, as God in all his wisdom gave us all something that we can access: Creation.

The season when I lay on my side in my garden to gain strength after intense treatments of chemo and radiation proved to be vital. Why did I do this? As gardeners, we know where to go when we need healing: Make the outdoors your prayer space.

So take you shoes off and walk on the garden – what do you feel?

2. WHAT TO PLANT?

STARTS

- Tomato
- Sweet potato (slips)
- Summer squash
- Pepper
- Eggplant
- Melon

- Lemon Balm
- Lemon Verbena
- Thyme
- Patchouli
- Stevia
- Lemongrass

SEEDS

- Cucumber
- Wild cosmos
- Red cardinal

EITHER SEEDS OR STARTS

- Okra seeds or starts
- Basil seeds or starts
- Hyacinth bean seeds or starts

3. WEED CONTROL

In March, it is windy and weed seeds come in from everywhere. If you have not found a way to control your weeds by now, you may feel like they are controlling you!

It is true that weeds take nutrients and water that your food crops want, but at times it is virtually impossible to remove them all, when as a beginner, you may have left many thinking they were veggie seeds coming up. Don't worry – wait for a great rain or a heavy watering and then pull the weeds that are strangling your plants. Several layers of newspaper can control the weeds that are in your path. Wet them to avoid the wind blowing them away, then put moist straw or mulch on top.

Sheet mulching with the kids

Your objective here is to make sure that not a stream of sunshine reaches your weeds. The cool thing about using this method of weed control is that over time, as little as six months, the mulch will break down and you can rotate this area to plant seeds. Furthermore, the paper will help hold moisture and bring in the red wiggler worms.

You thought that weeds were a problem, when actually they were your solution for where to plant your fall crops. Brilliant!

Another organic weed control solution is agriculture-grade vinegar. In May, the weather is plenty hot for vinegar to do the job. Beware, though, that nothing but hard work and lots of paper and mulch will choke out Bermuda grass.

Bermuda grass is welcomed by the chickens, so use this as a summer green to feed your flock. Chances are, weeding will feel better once you understand you are stacking your functions. If you do not have a flock, let the Bermuda dry out to a crisp before tossing in your compost pile, or throw it away!

A Mother's Day bouquet

4. WATERING

The simplest chores we tend to make the most difficult. The most-asked question I get from students and clients is: How often do I water? I simple say I do not know.

To know this answer would require that I know the slope of the garden, the quality of the soil, the root development of the plants needing water and that of those nearby… and the list goes on.

So, the only way to really know how much water your garden needs is to get in the dirt and check its moisture retention, along with the root development of your plants needing irrigation.

A general rule of thumb is that plants need the soil to dry out between waterings. Note that we are talking about the soil, not the plant itself. If the plant shows wilt and it is springtime, wait a day because it may rain. If it is summertime, do not wait too long, as the plant could burn. Burn starts on the tips where the fruit and blossoms are, so go ahead and give your garden a drink.

Ways to reduce water usage include adding cardboard and mulch to your garden, or to grow a summer cover crop, such as cowpeas, to hold moisture in.

Add compost on a regular basis so the roots of your plant stay cool and do not get as thirsty.

Take a class, or flip to the August chapter, and learn to implement passive water harvesting on your site!

Journal Space

June

Care less for your harvest than for how it is shared and your life will have meaning and your heart will have peace.

— *Kent Nerburn*

OVERVIEW

This month will feel like Christmas, with gifts of produce everywhere. For many, June marks the end of warm-season crops, but those skilled at the art of gardening in extreme heat will continue to raise food into July and August. Most of your tomatoes will stop fruiting until mid-September, as will lots of your other crops. The weather has just become too hot and dry to enjoy long days of gardening, so it may be time to take a break. If you have shade cloth with frames or a forest garden, then you will be able to extend your growing season.

JUNE CHECKLIST

- ☐ Plant mature, late-season veggie starts
- ☐ Harvest root veggies
- ☐ Store potatoes
- ☐ Preserve peaches, tomatoes, and other perishables
- ☐ Save your seeds!
- ☐ Sheet mulch your beds so they can start to rest

1. WHAT TO PLANT?

Plant mature, late-season veggies if you must. Soak them in compost tea or seaweed. Watch them closely until they are established, about one month.

One year I planted sweet potato slips, or starts, as late as the end of June using Hugelkultur (the ultimate raised garden beds) and had huge success. We harvested the large sweet potatoes in October following our first frost. The ducks enjoyed the sweet edible greens sweet potatoes are famous for. You should know that sweet potatoes are not related to the night-shade family, so the entire plant is edible, unlike traditional tuber potatoes like Yukon, Irish, and red potatoes.

Farm school students create a Hugelkultur bed as part of the day camp's permaculture learning adventures.

At this point in time try your luck with pepper plants, basil, okra, and eggplant. You can still plant zinnia seeds, purslane plants and seeds, and wild cosmos, especially the orange variety. You will be happy to find butterflies afloat after a long day!

2. THE HARVEST

For the most part, there is no science to gathering your homegrown veggies. Though it is best to pick fruit and flowers early in the morning during hot months, your schedule may not allow this. So, I think it best to do what works for you. But of course, the less time between picking and eating means a fresher and more nutritious meal.

Be brave about digging around and finding your treasures – it's such fun!

Harvest onions in late spring. Pull up a few onions to check their size. Once they're up, eat them; there's no putting them back. Or you can leave them in the dry ground, where these treasures can be stored for months to come and then can be easily found once the annual garden dies back. Discovering a previously hidden red Southern Belle onion is such a perk-me-up on a cloudy fall day!

Onions do well in a food dehydrator or an oven on 150-200° till dried. It takes the better part of the day or a nice warm windy day free of dust. In theory, a sealed air-tight jar should keep them for months, but even when dried, I store then in ziplock baggies or labeled canning-quality jars and they are eaten in our winter soups.

Pull up radishes. If left past prime, they will split open. If this happens,

just let them go to seed, then save these seeds for the next cool planting starting in late October. You need to let them dry out on the plant completely in order to germinate when sown. Or do what I do and treat the goats to some spicy red morsels with converted green tops.

Most of the root veggies store nicely left in the ground, as long as they don't sit in water. Onions, carrots, turnips, and sunchokes do well buried beneath the earth. If the plant is left too long, you may end up with some tough roots, which are great for stews or animal feed!

Potatoes store well in the ground for up to 3 months, while sweet potatoes will not be ready until the first frost. They do need to come out of the ground, though, if we are going to have a hard freeze.

Let some of your potatoes go to seed (sprout) in the pantry. We all do this without trying, but especially let your organic seed potatoes sprout, since the cost is a factor. Potatoes and garlic you buy in the store have usually been treated so that they do not sprout. That is why you need to start out your first time with specified seed potatoes and seed garlic.

The best storage for most veggies is in a place that is cool and dark year-round. Since this is hard to come by in our warm and humid climate, most summer veggies will be best enjoyed if picked and eaten right off the vine, or shortly thereafter. Even refrigerators are often too cold and moist to sustain your veggies' lovely summer flavor.

This is why I advocate canning, as well as old-fashioned sharing, since most everything is ready to harvest all at once, but all those wonderfully juicy peaches and tomatoes are not great keepers. It is okay to cut the rotten part off of your harvest for fresh eating, but don't can it or put in the refrigerator, for food safety reasons. There are already hundreds of great resources and books on canning; for this reason I have refrained from writing something that already exists. So I encourage you to obtain a good resource, invite a friend, and have fun putting up food, as they say!

3. SAVE YOUR SEEDS!

Most seeds should dry on the living plant to the point of being brown. Then once your seeds are completely dry, store them in airtight glass in a cool, dark place. Date your seeds and label well, as too often pumpkins get planted instead of butternut squash – and though surprises are fun, you may not want a pumpkin patch!

For seeds that come out of fruit like melons, cucumbers, eggplants, and etc, apply similar seed saving applications, but don't let the plant dry on the vine. Rather, choose the best fruit and cut it open, then remove the seeds and rinse off the remaining food particles. Lay the seeds on a screen to dry out completely.

Tomato seeds are the trickiest. I like to put them in a jar of distilled water and use the seeds that float. You will then need to dry the seeds. I have found that paper plates work best for drying my tomato seeds this way, as I can stack varieties easily and write information on the plates.

I have never gathered fruit seeds or pits from my trees, but I know that peaches are easy to

Canning Creates History

Canning soups, sauces and other delicious things is a tradition many folks love.

Harvest time was all summer long on my grandparents' ranch; with each explosion of crops came days of scalding ball jars and scorching the golden lids and rings. My grandmother would pack, but not over pack, the jars full of green beans, corn, tomatoes, and homemade sauces all summer long.

Essentially, she was putting up food for the winter, and that is just how she said it. I helped with simple tasks such as washing the green beans in the sink and then cutting them to jar length. I never lasted as long as my grandmother, who worked over steaming baths of water and a pressure cooker until it was time to prepare the evening meal. Thinking this was such hard work, I moaned about how tired I was, which prompted my grandmother to remind me of all the modern conveniences I enjoy like running water and electricity.

I loved to hear her stories of canning green beans when she lived in a logging camp with my grandfather. My grandmother was a city girl by nature, but hooked up with my grandfather, a John Wayne kind of a guy. She and the other ladies of the camp would make a whole day of canning; up in the woods with no electricity and running water, canning truly was work. But in those times, they worked for food or they didn't eat.

"Yes," she would say, "we had wood stoves that the men would fire up, but it was up to us ladies to keep it burning hot enough all day so the pots of water would keep at a constant boil."

They had plenty of firewood, since their beaus, as my Gramma called them, were woodsmen.

replant and grow by seed, as are blackberries from the vine. More than once, I have tossed a peach seed aside that has grown to be a lovely tree. The birds see to it that blackberries get scattered to become brambles of fruit for all to enjoy!

Seeds, when sealed and properly stored, have been known to keep for years and, in certain cases, centuries! However, I think it is best to use seeds the following season. Compost old seeds after two years. Though some may still germinate, just let them find a home in a pile of matter.

It was from those stoves that they heated their water to boiling, prepared their food, sterilized the jars, and canned fruits, veggies, fish, and meat.

My grandmother assured me that no one ever died of botulism.

My goodness! Risky business canning certain veggies and meats off a wood-burning stove.

But with ease, my grandmother said, "In those days there wasn't much fuss. if we came across a bad jar we just knew it the way it opened – a bad seal meant a bad meal, so don't eat it."

Simple, right? Even with our advancements, we have breakouts of food-borne illness all the time. Yet these folks simply relied on their good sense, commonly known as intuition.

These red skinned peaches and many other fruits make delicious jams and jellies.

4. PRE-FALL GARDENING

By June, most of your crops are spent, full of pests, and tired, so pull them up for the compost or let your chickens glean the remainder. Once they have cleaned, scratched, and added manure to the area, add fresh compost or sheet mulch and give it a rest.

After 4 to 6 weeks of letting the bed rest, if you want to have a fall and winter garden, add three more inches of compost on top of the sheet mulch and plant a cover crop of black-eyed peas. This will remind you to keep your bed moist while the peas are busy adding nitrogen to the soil and food for the table. Black-eyed peas love the heat and, as a ground cover, further nurse the soil for planting later in September and October. Even when the peas are growing, you can add compost so long as the soil is fully composted and you do

Trim and pull up the spent plants for compost.

not cover the plants. This will assure a ready fall garden.

Growing fall tomatoes is tricky and requires that you have a healthy and mature plant in August. Hopefully some of your spring tomatoes pulled through and will start putting fruit back on once it cools. Tomatoes are high feeders, so make sure you fertilize well with an organic source of refined compost high in nitrogen and full of life. My favorite fertilizers for tomatoes are fresh worm castings. They're the perfect compost for all plants!

Journal Space

Garden Plans

SUMMER

July

She said, "I want to increase the value of my life." I said, "Grow a garden, an edible one, and of course plant lots of flowers, too."

We dug up the front yard where the sun shines, then all the neighbors talked. "Oh, my" they said. "This is not a front yard garden."

"You're right," we said. "It is a garden to play in, and you may." They didn't, but we did, and the butterflies joined in our fun.

— Elizabeth Anna Samudio

OVERVIEW

Take a break in July, but keep up with the weeds, or at least keep them at bay. I love to put on my lightweight shorts and t-shirt and make a day of weed attacking. I will use a small sprinkle to cool off while I water the dry crops and loosen weeds from the ground; before you know it, I am covered with mud and digging every minute of it!

JULY CHECKLIST

- ☐ Take a break!
- ☐ Weed your beds occasionally so they don't get out of hand
- ☐ Observe and brainstorm new planting sites
- ☐ Remember to water the compost!

1. HOW TO WEED IN THE HEAT

Don't toss that Bermuda grass in your compost pile if you want it out of your beds. If you want to kill it, shake off the soil and lay it on concrete. It will take a few days to burn the life out of it, or if you have chickens, they love to pick at Bermuda salad.

Mr. Cline, a late dear client of mine, came by my urban garden one day and caught me drenched in muck. Laughing, he said, "You just need a good excuse to play in the mud, don't you?" I was found out!

If you prefer choking out weeds as a patience practice, try putting down layers of newspaper and cardboard. Bear in mind, though, that Bermuda is tough stuff, so keep at it. Even when I use the sheet mulch method, I find I have to weed the edges and create a barrier if my bed is next to Bermuda turf.

2. CHOOSING YOUR YEAR-ROUND EDIBLE PLANTING OPPORTUNITIES, AND CREATING MICROCLIMATES

A microclimate is the climate of a small area that is different from the area around it. It may be warmer or colder, wetter or drier, or more or less prone to frosts.

Microclimates may be quite small – a protected courtyard next to a building, for example, that is warmer than an exposed field nearby. Or a microclimate may be extensive – a band extending several miles inland from a large body of water that moderates temperatures.

Cornell Gardening Resources, http://www.gardening.cornell.edu/weather/microcli.html

When choosing a planting site, you need to find a sunny location for warm season crops with easy access so you do not get lazy about checking on their progress. The site will need to get eight hours or more of full sun a day for annual veggie crops. Also consider sustaining your garden and food crops with natural rainfall. We'll talk about that more in August.

If your space isn't a square shape, don't get hung up over it. But plan accordingly. If this means removing some shrubs, just do it. If it means removing a tree, then make good use of the wood. Use it for firewood, swale debris, or rent a wood chipper to make some mulch.

Consider pets, children, and vandals when choosing your location. Thieves seem to prefer above-ground crops. Make the front yard garden a place where you grow root crops if you have food vandals. In truth, the vandals are usually troubled kids, since most homeless folks have access to food. And it is good to share a portion with neighbors or those

Planting sweet potato slips with farm school kids.

Victoria and Mary Magdalene, My First Texas Gardens

When we moved to our home in Fort Worth, the first thing we did was put in a veggie garden — even before we stripped the floors, painted the surfaces, and fixed all the cracks in the walls. Ahh, the best spot and space that called for life was in the front yard, between four trees the previous owner had planted like guards. Sorry to say, but I later had to remove one of the solemn soldiers to make room for the sun to shine. I didn't give it a second thought that our main edible garden was smack in the middle of the front yard. Like, duh, here is the sun and a big empty spot — where else do you plant melons, cucumbers, and tomatoes?

To my utter dismay, a guest of our home group one evening said to me, "Here in Texas we keep the front yard a yard. It's not acceptable to put a big mess where all the neighbors can see." My mouth dropped, but I still went about doing the work of planting an edible garden where all could see, and we played, learned, and busted our butts to overcome the existing black clay.

As you might suspect, the neighbors did talk. The city code enforcement folks sent me lots of letters, as certain neighbors had a hard time with me growing food in my front yard, but by now, I was growing food on our city site and also teaching others to find the best spot to create a garden to eat from and grow with.

One day an elderly lady nearing ninety years of age came by to let me know I had a Victory Garden. Some of you may know the history behind that name. During World War I or II, the government of the US encouraged housewives to grow crops in their yard to make up for all the food that needed to be sent overseas to our soldiers. Over 40% of the food consumed during that era was grown in ladies' front and back yards. This patriotic movement among women changed the face of home gardens from decorative to edible and granted the US some much-needed food security.

And so Victoria was my garden's first name, until I got sick and redesigned my agricultural site to be a sacred spaced filled with plants, scents, and a water feature. Mary Magdalene, a reformed harlot in the Bible, was scorned by the religious zealots of that day, yet esteemed publicly by Jesus and gave her love offering by pouring expensive pure oils on her Lord's feet. Thus, Mary Magdalene, one of my favorite characters of the Bible, also became the name of my healing garden--where much peace, rest and renewal grew.

who are passing by.

Also consider using existing fences as a border to create an edge garden. This is where you can grow berries, grapes, pole beans, melons, and cherry tomatoes.

Check to see where the sun rises and sets, then use this as a guide when determining what to plant. For example, leafy greens and cilantro do much better in cooler spots, and likewise, you may take advantage of the hot afternoon sun with crops like okra or eggplant. If you have a lot of west sun, then create some shade in the summer months by planting a tree that loses it leaves in the fall. I have a lot of west sun, so I have planted some princess trees along the west to nurse the other plants during summer's blaze.

Fences can be screens that shade most of the garden, or they can keep the summer breeze from flowing. Remember that here in the northern hemisphere, it is the south and north vortex that cool us in the summer months. Bear in mind that most plants do best with a southern exposure and that if you have a solid wood fence blocking it, you will need to create your own solution. Drill some holes in the wood so that you will still have privacy, yet also gain the airflow and sunlight necessary for crops that do best with morning sun. A man I just visited loved this idea and plans to make good-sized holes in his solid wood fence and then will install bamboo plugs in some of the holes as an accent.

You may want to put a winter garden on your driveway in a large, moveable container. Often, our hard surfaces can become miniature microclimates, allowing you to garden year-round, but we'll talk more about that on the next page. If you are from a cool region like me, cool weather gardening is welcomed. No heat, no bugs, hot coffee – um, count me in!

Another factor to consider is the direction of prevailing winds. Since I do farm in the winter, I make sure and protect my crops from the northwest winds, yet I welcome the southern breezes of the summer.

> **Since it is dry this time of year, be sure to water the compost. Put a vapor barrier on top to trap moisture in. Consider reusing large plastic bags taped together with duct tape. Soil and feed bags also work great, so ask your neighbors that buy bagged compost if they have any left over. Make sure to take the barrier off in the event of rain.**

Remember, a living garden is not set in stone, but rather, grows in the earth and can be renewed time and time again. Even permanent crops are transplanted or taken out for the best yield and landscape health.

Given this out-of-the-box criteria, I hope that when it comes time to choose a space, you will get creative! Perhaps try a sandbox container that in the summer can be a place for your children to make mud cakes, but in the early spring can become a potato patch. Add plenty of compost because potatoes like a sandy loam; make sure and don't make the mistake of planting them in sand alone. And wow! You are a potato farmer. By all means, when it comes time to harvest, encourage your young ones to dig for yukon gold!

3. KIDS' CORNER

We all remember mud pies in the summer. By now you have gotten dirt under your nails, and if not, it is about time you men and women start to play!

This is a real crowd-pleaser among children and takes so little effort. Your kids will love this and if the neighbor's children catch sight, they will join in the freedom.

Here's what you'll need to build great mud pies!

+ Buckets.

+ Dirt

+ Grass or weeds

+ Worms, optional

+ Water

+ Seeds (past date so as not to waste expensive seeds) or legumes, as they give quick results.

+ Creativity! (Hint: make the pies in layers in small containers and then sprinkle with seeds that may sprout.

Farm School isn't all work and no play!

From Elizabeth Anna's Farm Kitchen: Texas Summer Tapenade

Tapenade, a French provincial food sensation, has found its way to Texas tables. Using our fresh, local, in-season fruits and veggies, summer small meals are a breeze!

Ingredients

- Three ripe tomatoes
- Two garlic cloves, to taste
- Five basil leaves
- Three tablespoons of red onion
- One red or green pepper
- Chopped jalapeño (only use a pinch at first, then adjust to your taste)
- 10 pecans
- ½ cup green and/or black olives, preferably fresh
- 5 tablespoons of olive oil

Directions

1. Toss all ingredients together.
2. Cook on a roasting pan or cookie sheet for 15 minutes at 400°, or over fire on a grill.
3. Let cool for at least 20 minutes.
4. Mix in a blender or small food processor.
5. Spread on fresh bread, pita, or plain crackers.

Variations

- For a taste sensation, add goat cheese!
- Use tomatoes and peaches with fresh mint leaves and omit the olives and basil.
- Make it with just peppers, nuts, olives, and oil for a simple fire-roasted dip, served with blue corn chips.
- Add roasted summer squash!
- As long as you have oil and herbs on hand, have fun and get creative with the recipe!

Journal Space

August

Deep Summer is when laziness finds respectability.

— Sam Keen

OVERVIEW

This is the month to make sure all your soils are covered with either a cover crop or mulch. Mulch can be leaves, straw, hardwood, paper, or even natural fibers. If you have irrigation, you need to check for breaks and over-spray. This month your garden needs a vacation, and those fall crops you may have started early on need tender care.

AUGUST CHECKLIST

☐ Make comfrey tea to relieve your plants of stress

☐ Hire a farm-sitter and go on vacation!

1. STRESS SOLUTIONS

It's best not to plant this month, but treat your existing plants to compost tea or a tea made with comfrey leaves. This nice sun tea will help your prized plants get through the stress of summer.

This is also a good month to make yourself and your friends a batch of holy basil sun tea. Once brewed, ice and serve with a lemon wedge – it's refreshing, a great natural antidepressant, and a superb antioxidant.

Comfrey Tea For Your Plants

1. Fill a large glass jar with comfrey leaves and/or roots.

2. Fill it with filtered water.

3. Set it out in the sun and let it brew all day!

2. PASSIVE WATER HARVESTING

Passive water harvesting is, without a doubt, the most important element of growing food in the future – and the future is now.

Whenever I mention this to my clients, they usually think I am talking about a rain barrel. This is not passive, since it takes effort to get the water to the plant. When you have a passive system in place, you do nothing; the earth and the plants do the work.

Cut on contour to form a basin, or swale.

Use the dug-out soil to build a berm on the lowside.

Results in absorption and decreases run-off = Passive water harvesting!

Don't be fooled: there is work in the beginning, but once you have set up this system found in nature, you will be amazed at the benefits. Once you have done the earthwork and planted your crops, you will save money, your plants will get the best water (rain!), and you will save time by not having to water as often.

To begin with, you will need to find the contour of your site. Contour lines basically run level across a slope. This will be pretty easy if you have a slope, but land that is seemingly flat will take more skill to find the contour. The contour lines are the areas that should be worked, by cutting out swales to hold moisture in the ground.

The lines in this simple diagram show the level or contour of the ground, not to be confused with the grade, or slope.

If you are not so blessed to be on a slope, then you will have to do some work using a laser or a simple A-frame tool to mark the contours on your land. Another trick is to wait until you have a good rain, and then observe where the water settles and how the low points are positioned. If I learned anything by getting certified in permaculture, it is the genius of earthwork on contour for passive rainwater harvest. Needless to say, you do not need your certification to learn this basic principle. However, we offer classes through our shop in Fort Worth (check elizabethanna. net), and I strongly recommend that you take advantage of instruction on passive

Lines and wooden stakes mark the lines to dig the swales.

water harvesting. If you do not have time to get certified, Youtube has some great examples of how to find contour on your site.

Once you find the contour, you are ready to mark your lines, dig your swales, and form the berms. The swales will hold water and the berms will help the water to slow down and to flow into the desired location, thus resulting in absorption and percolation of the water into the

Vacations and Discoveries

If there ever is a time to take a holiday from the farm and garden in North Texas, it is in August.

This is the time we get away, and it is also the time we lose plants and sometimes a chicken or two. But the time away is needed.

But in fact, the largest project I have worked on to date was retained to be designed in August. This was not your average job, but rather three acres that sprawled over an earth bowel formation with a slope coming from every direction. Everywhere I walked, there was erosion and the evidence of past flooding throughout the site, all the way to the back door. I was called on for my water-saving expertise. I must admit this problem was a bit daunting, yet considering all I have learned about earth and water absorption principles, I knew I was on the brink of a great opportunity to make permaculture principals shine.

While in the lake waters of the northwest, I ran my hands through the sandy mud. I felt a foreign object. I tugged till I unearthed a piece of wood — not just any piece of wood, but an inspiration of art created by some type of worm or insect, fashioned like a blueprint. It was on this find that I based my permaculture design.

The client and I were both so happy to find that after our earthworks were in place, 97% of the water was absorbed on the 12-15% slope, and all flooding was eradicated from their home!

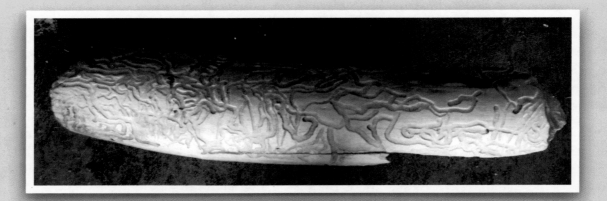

earth.

This earthwork seasons in time, so to speak, and will provide more planting opportunities over the course of a few years, rather than right after its formation. The main function in the first year is to store water in the ground, creating absorption and starting the process of sustaining your plant growth with outside energy. What you are planting will determine where to plant the first year. Even in the first year, your crops of trees, etc, will benefit, and so will you financially.

Evidence of the power of a passive water system on a contoured hillside

3. SAVE THE BEES!

I have fond memories of large white clover in all the tall weedy lawns I played in as a girl. In the spring my shoes came off as soon as the ground warmed up, or rather, thawed out. I made a game of running in the soft blades without stepping on a honeybee. I got stung often, but it didn't stop me from experiencing the pleasures of what were then called lawns and are now called meadows.

In the sixties, clover was actually added to grass seed to improve lawn conditions, since this leguminous plant fixes nitrogen. But to our own demise, chemical lawn companies and weed killers wiped this jewel out, along with other longstanding plants that once provided food for bees. Unfortunately, we still live in a turf culture obsessed with the monoculture of blades.

Most of us know that bees are in trouble. But the multitudes do nothing about it. You, however, are not like most. Let me congratulate you now for being part of the solution to saving the bees and our planet. Surprisingly enough, it is so easy for the flora to remediate and, in the process, create a banquet for the bees.

If you do, however, have a lawn that has been compromised by poisons and artificial fertilizers, back off of all weed killers and start using a variety of composts to feed the earth your lawn grows on. You will not only help bring back a healthy place to play, but you will also save lots of water through natural lawn care! If you want to speed the process up, add aerated compost or comfrey tea (recipe can be found at the beginning of this chapter), beneficial nematodes, and then broadcast clover seeds in the cool of Autumn.

Join thousands of us working to save the bees by throwing seeds about to naturalize our surroundings. For better success, toss the seedheads where you find existing dry clover heads. For us in North Texas, the Native American Seed Company (seedsource.com) carries a white prairie seed that naturalized here and is perhaps native to our prairies.

Find key spots to broadcast. Remember to observe how this plant grows in the wild.

OTHER BEE-LOVING NATURALIZING PLANTS

- ✦ Dandelions
- ✦ Mustard
- ✦ Arugula
- ✦ Basil
- ✦ Mint

- ✦ Thyme
- ✦ Passion vine
- ✦ Fruit trees
- ✦ Bee balm

SAVING THE BEES YEAR-ROUND

To seriously help the bees, stagger flowering plants so they have access to food most of the year. Besides food, make sure they have water to drink.

A small list of plants for all four seasons to nurture the bees:

BEE-LOVING PLANTS FOR WINTER AND EARLY SPRING

- ✦ Fruit trees
- ✦ Carolina Jasmine
- ✦ Dandelions
- ✦ Wild Arugula
- ✦ Wood Violet
- ✦ Primrose
- ✦ Forsythia
- ✦ Lavender

- ✦ Quince
- ✦ Redbud trees
- ✦ Scabiosa
- ✦ Crocus
- ✦ Sweet peas
- ✦ Honeysuckle
- ✦ Marigold

BEE-LOVING PLANTS FOR LATE SPRING AND EARLY SUMMER

This is the easiest season to find blooming plants.

- Bee Balm
- Most herbs
- Roses
- Daisies
- Salvia Greggi
- Black and blue Salvia
- Hollyhock
- Tall phlox
- Coneflower
- Day Lily
- Canna Lily

BEE-LOVING PLANTS FOR LATE SUMMER AND EARLY FALL

- Black-eyed Susan
- Morning Glory vine
- Cypress vine
- Woody herbs
- Passion vine
- Roses
- Vitex tree
- Rosemary
- Russian Sage
- Mexican Marigold

BEE-LOVING PLANTS FOR LATE FALL

- Hops
- Passion vine
- Hummingbird Mint
- Apple Mint
- Asters
- Blue Basil
- Roses
- Marigold
- Most Salvias

Beekeeping is more than a hobby, but rather a calling. If you feel the noble tug, I highly recommend that you go through a masters program for training.

4. WATER THE ANIMALS!

Make sure if you take a vacation that you hire and pay someone a fair wage to keep your plants and more importantly, your animals, well hydrated. During the dog days of summer, we change our pets' water twice a day, allow the fowl full-range to find the coolest spots, and make sure everyone has plenty of shade. We spoil the dogs and let them enjoy fans and air conditioning.

One of our geese cooling off in the pool

Journal Space

September

The first supermarket supposedly appeared on the American landscape in 1946. That is not very long ago. Until then, where was all the food? Dear folks, the food was in homes, gardens, local fields, and forests. It was near kitchens, near tables, near bedsides. It was in the pantry, the cellar, the backyard.

— Joel Salatin, author and owner of Polyface Farm

OVERVIEW

The red cardinals have come back to feast on the sunflowers; they remind me of love, steady and pure. The end of summer is a return to routine, tradition, and hope of cooler days. But more importantly, it's when the night and day make a truce and the balance of creation is cast, granting us with the autumn equinox.

Considering the seasoned farmer I would grow into, it was of no consequence that I married my husband, who I can proudly say I am still with after 25 years, on an important agrarian day: the autumn equinox. It takes place when the plane of the earth passes through the sun, somewhere around September 21st to the 23rd, depending on the year. Aside from the spring equinox,

> **SEPTEMBER CHECKLIST**
> - ☐ Harvest summer crops
> - ☐ Save seeds
> - ☐ Prepare for fall and winter gardening, or allow for restoration
> - ☐ Order garlic and wild-flower seeds
> - ☐ Choose the right trees for your site
> - ☐ Consider adding livestock to your farm

these are the only two accounts when this special astronomical occurrence takes place. The autumn equinox marks a shift in the long growing season and gives farmers, hunters, and gatherers an alert to unearth the crops and celebrate God's bounty. Harvesting has a romantic mysticism around it and, as beautiful as it is, like marriage, it is hard work. Whether it is hot, raining, or windy, the work must be done. But like relationships that are rooted in commitment, you will in turn get to enjoy the fruits of your labor. So invite friends and dig in, or be ready to put in long hours.

Elizabeth and James' wedding day,
September 23, 1990

1. PREPARING FOR FALL AND WINTER GARDENING, WHILE HARVESTING SUMMER'S BOUNTY

Many places, including North Texas, have overlapping growing seasons. Unless you have a lot of space, you will need to harvest your summer crops to make room for your fall crops. If you are like me, you will have lots of green tomatoes that stay hanging on the vine as the days get shorter and the chance of a frost get closer. Whether I leave the tomato vines in the ground or not just depends. Sometimes I will keep the fruit on the vines and hope for sunny days to turn them more of a light green. And fried green tomatoes are a favorite of many chefs, after all. But if I want their space and am tired of looking at them, I relish in pulling them up. I take the green tomatoes off the vine and then place them in paper sacks in a cool dark place to further ripen, and later this versatile fruit will join my peppers in a sweet, hot chutney. With Thanksgiving right around the corner, this preserve will come in handy. But sometimes I am just lazy and feed the rest to the goats, who appreciate Summer's leftovers.

Since I love to garden in a bug-free zone, which just so happens in the winter, I go ahead and dig up the rest of the nightshade potatoes, but leave the sweet potatoes in until after the first frost. I harvest the rest of the melons and let the animals eat the ones that may never really mature. I start cutting the basil, making pesto to put up and also bundle some to dry out. Sometimes I pull some of my mint to dry out too, but most of it winters over pretty well so we have fresh mint year-round.

Though I know that early morning is the best time to gather the harvest, sometimes I do it under the moon once the day's work is done. By September I have tired of okra, so I take my goats in on a leash for a selective gleaning of the garden and let them nibble away. I do leave my peppers in that, by now in the fall, are glorious, especially the red cayenne and small sweet peppers. There is no shame in waiting to keep the goods growing as long as you can. Sometimes I have even transplanted some of my warm-season plants into pots and brought them in the greenhouse. It is hard to let all the summer's flavor come off the vine, but overall I have learned to let go and prepare for winter by harvesting in due time.

SAVE YOUR SEEDS

Part of celebrating the harvest is holding back a portion of seed and feed to give back to the community and to next year's or next season's planting. By fall, most of your plants will have lovely seed heads. Remember that the seed needs to completely dry on the plant and seemingly die to reproduce. I love the parable of the death of a seed: at first it seems like a dream that comes to naught, but then out of nowhere, that which we held onto and then had to let go of suddenly emerges and manifests in our lives. Jesus talked in parables because he knew supple hearts would hear mysteries hidden in dark places. The bible sometimes reminds me of Where's Waldo? books. It is not the obvious that we learn from, but the hidden things that come alive, and so it is with seeds. Once they die and fall to the ground, the life buried inside springs forth, producing even more seeds! Food security at its finest. I especially like to let some of my best carrots go to seed since the bloom is so beautiful, appearing like a sea creature. If all the seeds are not dry by the end of September, you can tie a paper bag tightly around the seeded flower head so the dry seeds are in captured in the bag. Some seeds, like lettuces, are so tiny that I often just don't let the seed dry, but rather put them in a jar with stem and all and let the seeds fall out of the shells by themselves.

By now you will have discovered that most seeds have an outer shell that protects them from blowing and scattering all at once. Seeds naturally will float on the wind to places of opportunity and start a new colony. In a forest garden, this should be encouraged so that you are managing the growth and harvest, yet also allowing a natural system to develop. Less work for the farmer!

Fruit seeds are on the inside, so they obviously need to be harvested when the fruit is ripe. Save the seeds from your best melons, tomatoes, cucumbers, and etcetera. If you have planted a hybrid, several varieties will reseed. Keep in mind that GMO plants don't reseed, which is a good reason to grow more heirloom crops and open pollinators. Large corporations may want to own the seeds, but with heirlooms floating about like angels, on a mission to feed the meek, we strengthen food security and create healthy, interdependent communities.

EITHER MULCH YOUR BEDS, OR GET TO WEEDING

September is a great month to decide what crops, beds, or plots need a rest. Mulch existing beds that you do not plan on seed planting, or use green mulch, such as large-leaved plants, to cover the soil. This time of year, you should have plenty of greenmatter biomass in the form of sunflowers leaves, cannas, or princess tree leaves to sheet mulch with. But just so you are not disappointed, this will not kill existing Bermuda grass. No, after all my years of gardening, I've learned that a sharp hand tool, elbow endurance, and moist earth is the best way to get rid of this survivor. Chickens or herds that you are raising are also a great enemy of Bermuda.

If you decide you want to just let your garden rest, lay plenty of mulch, straw, leaves, etcetera, and keep it all moist so that the worms and beneficial nematodes stay alive. You will still get some weeds because the ground wants to cover itself, so you can weed-whack or pull them up and then toss them on the soil to dry out and give back what they have taken nutrient-wise from the soil.

But if you are like me, you may love to garden in the fall and winter, so September is time to get to work preparing your beds. I wait for a good rain to saturate the earth and soften the ground. Then I invite friends to lend a hand and make sure I have good food and drink to provide for them. I let it be known that I will feed them well with a good farm-to-table dinner. I start a little earlier than my guests arrive and first clean and fill the duck ponds up with fresh water from the hose. Sometimes I even add mint leaves to give myself a lift when I take a dip to cool off! Once everyone arrives, I am usually already wet and muddy from head to toe. This is not a job for the modest – so throw off your white gloves and dig in! Others will follow and many hands will make for light work. Make sure you have plenty of iced water and tea, saving the beer and wine for the end. With a party of helpers, we also end up with a few days worth of weed meals for my flocks and small herd.

If you do not want a party of weed junkies at your house, then take a few early nights to do this task on your own. You will be amazed at how lovely the evenings are when you dress light and take occasional dips to cool down.

Weeding is the most laborious part of getting beds ready, or perhaps you have flocks to let out and eat the weeds and the end of your warm season crops. You also may decide to sheet mulch over what you have, but if you plan on planting for a fall and winter harvest, you have got to get the ground ready. Once the weeds are out, treat the process much like you would in the spring. Fertilize with lots of compost and make sure if you are planting seeds that the top of the soil is fine and free of debris, so the tiny seeds can avoid fighting their way through hard clumps or masses of carbon, such as dried mulch. A lot of the preparation you are doing now will be for crops that may not be planted until October, so take your time preparing and building life back into your soils for a great fall-into-winter yield.

Remember that healthy soil and diversity is the best and least expensive pest control. More than just being organic, it is important that your soil be biodynamic, which simply means "alive." If your soil is nutritional, then so will be your food, and with denser nutrition in the food, you actually eat less. It's the best health insurance available! It is great to make your garden your primary health care source, full of fruits, veggies, herbs, and nuts.

2. WHAT TO PLANT?

There is no conflict with planting a cover crop and a fall garden at the same time. When it comes time to plant broccoli and cabbage starts, just harvest some of your peas, as it will be about that time anyway. Even if you want to plant seeds in late September, just pull the vines out of the way, make a furrow, and fill it with fresh-screened compost for root crops like carrots and beets. Bear in mind that the soil needs to get down to 60° for your carrots and other cool-season crops to germinate (grow). So if after two weeks, you still do not see bits of green coming up, replant or wait until evening for the temperature to cool down.

SEEDS

- ✦ Russian kale

- ✦ Malabar spinach

- ✦ Pole beans

- ✦ Winter squash – with backup protection into late October

- ✦ Pie pumpkins (or starts) – with protection through the winter

- ✦ Dill, flat parsley, cilantro, and nasturtium

- ✦ Black-eyed peas – as a cover crop to give back to the soil

- ✦ Savoyed spinach – late in the month when the ground is cool. This spinach with a triangular leaf is more heat tolerant and less likely to bolt during warm days.

- ✦ Lettuce – mid to late September

- ✦ Irish or early potatoes – choose ones with sprouted eyes for a head start

- ✦ Beets, turnips, and carrots – late September

- ✦ Rainbow chard – the plain green variety is much more cold hardy if you do not like to mess with cold frames or coverings in case of low temperatures.

Rainbow chard

STARTS

- ✦ Okra – early September

- ✦ Fall chrysanthemums – much hardier than marigolds and produce all the same benefits. Plus goats love to trim them back when they are finished blooming, which will help them bloom again in spring.

- ✦ Johnny Jump Ups or violas – late September. Plant under trees, as they tolerate some shade, unlike pansies, that need full sun.

September is a good time to plant black-eyed peas, as they assist in squeezing out Bermuda and other less desired crops, while restoring the soil at the same time. Not only will you have fresh black-eyed peas to eat, but this versatile and heat-loving legume plant will add nitrogen biodynamically. Though I have said it before, it is worth repeating that legume plants have nodules in their roots that, in layman's terms, pull nitrogen out of the atmosphere and add it to the soil! By planting cowpeas, you are stacking functions: food production, ground cover, choking out of less desired plants, and garden restoration. I am sure there are more functions that this amazing little bean plant provides, but these mentioned are reason enough for me to plant lots of them everywhere! And my goats love the dried up ones I let stay on the vine.

Elizabeth Anna's storefront in September

If you want pumpkins and winter squash, try to find starts, but seeds will also do just fine as long as you prepare to cover them in the event of a frost.

In North Texas, let the cool-season squashes stay on the vine until the first frost, usually in late October, to sweeten your crops. The thin, once vibrant vines, now like old shoe strings, will let you know that it is time to gather all that you have grown, except for a few potatoes that are buried deep in the supple earth, as well as sunchokes, which keep best in the ground. Like bunnies, they burrow down naturally.

Even though September is usually too early to plant fall seed garlic, if you wait until October to purchase them, the grower may be sold out. Order seed garlic and also wildflower seeds, such as bluebonnets, wild poppies, and larkspur, to plant in October. The flowers need to germinate through the cool months and will be a wonderful variety of bee pollinators.

When selecting garlic to plant, make sure the bulbs are large and firm. To plant, take the bulb apart and plant the clove root, with the hairy part just below the ground.

Raised beds don't work well for growing garlic, since they dry out in plains climates. Instead, plant in good, deep soil, 5 or 6 inches apart.

The same goes for seed potatoes. Cut the seeds into thirds and plant directly into the ground with the sprouts up and the white, cut part of the potato in the ground. As potatoes grow, keep adding compost and they will keep producing until they freeze. For September planting, you may need to dig 8 inches down to find cooler ground, as they generally will not grow in soil above 60°. Potatoes reach maturity after 90 days, but after 60 days you will have baby potatoes. If there is a frost, cover the area with frost cloth or straw and you are good! Planting the potatoes deep will also help to protect from a frozen ruin.

3. CHOOSING THE RIGHT TREES FOR YOUR SITE

If you are developing a forest garden, also known as a food forest, you may want to do your research now, since October is one of the best months to plant many trees in North Texas and other plains and cross-timber regions.

To begin with, there are many questions you can ask yourself to narrow the selection down. I don't know about you, but for me, choosing a tree is much like being at the fork in the road. One way says Silver Lake and the other says Nooksack River, and all you have is an hour left to explore. You want to make it count. The lists of questions I have compiled are sure to help you make a thoughtful decision. By the way, I chose the river route to gather stones fashioned by the rushing of water.

We need to do some intentional tree pondering. As you do this fun exercise, study and take lots of notes, drive around, and see what trees are thriving in areas that look like your site, or mimic a site to get the timber and edible species you desire.

✦ How many functions and/or purposes does my tree have?

✦ Is my tree edible? And if so, what parts?

✦ Can I get away with one fruit tree (an open or self-pollinator) or will I need two variations of the same species?

✦ What kind of trees do your neighbors have? If you notice peach trees all over the street, then you may want to choose one too. Get to know the neighborhood, as pollinating bees can travel within a 5 mile radius or so and it may be beneficial to choose a tree that will have nearby companions.

✦ What are the water requirements for my tree?

✦ Is my tree drought-tolerant, or will it benefit from having a taproot tree planted nearby to nurse it?

✦ Does my tree provide shade from the west sun in the summer but still allow light in during the winter months? Princess trees are great for this!

✦ Does my tree provide fodder for any of my stock, including honey bees or wild birds?

✦ Does my tree have abundant biomass for cleaning the air and sheet mulching (laying green matter on the ground to trap moisture in without adding more carbon)?

✦ Is my tree hardy to recorded temperatures or do I need to protect it from the winter prevailing winds?

✦ Does my tree need to be a winter evergreen wind block?

✦ Does my tree need to create or hinder a vortex from the north and south summer prevailing winds?

✦ Does my tree reproduce naturally, providing me with some extra income?

- ✦ Does my tree add nitrogen to the soil? Is it leguminous?
- ✦ Is my site best for species trees or hybrids?
- ✦ Does my tree provide fuel, oil, or timber?
- ✦ And last but not least, does my tree give me pleasure?

To help in case you are overwhelmed, let me give you an example of how I would choose a tree in the development of my forest garden.

To begin with, I need something to draw the water up from the depths during drought, but I also want something that will shade my home from the west sun in the summer while letting the sun come through in the winter. Since I live on a busy street, it would be nice if the tree has large leaves to clean the air, and also to become food for my goats, as they eat a lot and I would rather grow their feed than buy it. I also have a small budget, so the tree I choose needs to grow fast. I know that eventually I would like to add fruit trees, so I need a tree that attracts pollinators. Furthermore, in the summer months, I would like to use the large green leaves as green mulch to help retain moisture in the soil. As a bonus, I would like for any parts of the tree to be edible for human consumption. And last but not least, I want my tree to be beautiful.

Pawpaw trees will bloom in the spring.

Not a lot of trees have this many functions, but the princess tree, also known as Paulownia, does – and so my choice is made. This Asian native shall make its mark, tall and proud for all to come and ask, "What kind of tree is that?"

About species trees and hybrids: generally speaking, species trees are those found in nature and not cross-bred by people to provide a larger fruit or bloom. For example, a Mexican plum is a species tree native to North Texas and beyond that provides small tart plums that are great for sauces and juice. I have found that it is important, if you have the space, to provide some species trees for the native wildlife, mixed in with domesticated trees for the birds to enjoy sweet-fleshed fruit in the hot summer months and then the native pantry of the fall. Squirrels, too, who will want to take and bury some of the fruit and nuts for the colder months to come.

SPECIES TREES, FOR BIRDS AND PEOPLE ALIKE TO ENJOY

+ Native persimmons — small trees that look a bit like privet trees, but produce caramel-colored tiny fruits full of seeds.

+ Mexican plum

+ May haw

+ Native pecans

+ Russian hardy pomegranates

+ And to go with your trees, long vines with berries, such as wild mustang grapes

Basically, these fruit-bearing plants take little care, less water, are less likely to get bug infestations, and therefore are important to weave into your food forest. You are now beginning to create a tree guild, which with several guilds combined, becomes a miraculous forest of food living in harmony!

Grow greens in the fall and winter under your fruit trees!

PLANTING RAIN AROUND YOUR TREES

Dig around your fruit trees and "plant" rotten wood and compost to hold rain.

Some of my favorite trees need more water than the drought-tolerant varieties, so I will help them out by doing what I call "planting rain," which is adding dead, sponge-like wood into the tree hole, as well as creating a moon or half-moon berm, depending on the direction of rain flow.

Here's some expanded detail for the novice: Create a half-moon around your tree to increase water absorption and decrease run-off, making sure that the half-moon lays across the grade, or slope. If your site is fairly flat, then make a full circle around your tree, known as a tree well. Use the existing soil from the hole you dug, whether it is rocky, sandy, or laden with clay, and make a half-moon-shaped berm on the bottom of the slope where your tree is installed. The berm should be out about 2 or 3 feet from the tree, depending on the size of the plant and how much soil you dug out. You can widen the size of the hole that the tree is going into, but still only plant as deep as the root crown. Bury rotten wood along the sides of the hole instead of in the bottom, since decomposing wood will make the tree sink and it's not healthy for the crown to be below the surface. Fill in the remaining space with good compost to give your tree a happy home and a nice environment for years to come.

I know this all seems like overkill and these simple practices seem to throw a lot of new gardeners, but in this case, more information is better than less when creating a food forest.

DETERRING BIRDS AND SQUIRRELS

You may be thinking that if you offer fruit to the birds and squirrels that they will come to dine on the good, along with the fair fodder that you don't mind giving up. You are absolutely right, but birds are very smart and can be easily trained, if you will take the time once the fruit starts to ripen. You will be using Dr. Skinner's conditioned response psychology, or modified behavior training, and it is best to choose a discreet place to be. When the birds eat from the species trees, play some music or run some water, as they love the sound of both. If you are a musician, they will over time come back to hear you play melodic sounds. In doing so, you are training them that when they eat from certain trees, they get rewarded. It would be great if this in itself were enough, but the birds do need to know that undesired behavior, such as taking a huge peck out of a fat, juicy, red-skinned peach, is not okay. You will then need to discipline them with scare tactics. A simple pellet gun or even a loud pop gun or firecracker will work. Pop! goes the weasel and the birds will flee. Birds are very social, so you will not have to warn every single bird; they will do that for you. But you will have to play this game more than once for success. The more you condition the birds, the more manners they learn. They are doing you a great favor by hanging in the trees, anyway. They are eating pests and releasing slow fertilizer from the trees. Besides that, they may be planting a seedling you will later want in your forest.

To aid even more in training, hang some nice birdhouses and give the birds treats. This is a great homeschool project and a good way to learn about our native and migratory birds. Beware though, white doves can get a little aggressive with the food, so skip some days now and then. Make sure to plant plenty of sunflowers for the fat doves, along with giving them seed mixtures. If you keep cats, the birdhouses need to be high up or the cats will make game out of your sweet cooing friends.

Squirrels are another story, as they are very territorial critters. Try Skinnerian training on them or do what some who are living off the land do: once the small rodent is fat, enjoy the meat. My husband, who grew up in the woods, loved to hunt small wild game as a young boy and held to the rule: if you kill it, you clean and eat it. Squirrel, according to James, is delicious. So don't be a resentful peach tree grower; if the squirrels set up residency, create some boundaries.

4. CONSIDER ADDING FARM ANIMALS TO YOUR SITE

Stock, or farm pets, as some call them, will turn your one-dimensional farm into a poly-culture, rendering a multiplicity of functions and increasing yield! But the decision to start raising flocks and herds should not be taken lightly, as they need to be near your home for observation and protection. For instance, it is not okay to get the stock and then make the coop. That is back assward; pardon my slang, but foul are someone's prey and you are their protector. Even some of the best hen houses can be invaded by raccoons, who stick their little paws in, pull the hens' heads off, and then suck their blood. Lovely, right? Not! Instead, I want to paint a picture that will make sure your stock is happy, healthy, and protected.

A small flock increases yield in the form of a poly-culture.

About twenty years ago, we bought a house in suburbia on a nice big double lot. We looked in the poultry ordinances so as to not step on anyone's toes. But as you can imagine, most of our neighbors still had a difficult time accepting the changes we brought to the southside of Fort Worth. Our lot afforded us the space to build our chicken coop fifty feet away, the legal requirement, from any building, including ours. So when the neighbors did complain and a code compliance officer paid us a visit, we pulled out a measuring tape and were found innocent. But they still complained about Jack, our sweet rooster, who then became dinner once he started his crowing call. We made a compromise in the hope of creating community in our small neighborhood. Soon enough, fresh eggs and homemade bread softened their hearts more to our ways, which were foreign to them.

It was in those early years of raising chickens and homeschooling that I began to think that just a few more would be a great project and more fresh food. My husband, who was not as agreeable, not only gave into my request for three chicks, but instead brought home twelve little fluff balls one early spring night. They imprinted on me, so like a mother hen, I always found them underfoot. When I would sit down they landed on my shoulders and made themselves at home, but their favorite time was when I would garden. We all, including the children, had

A Goat Memoir

Having had so many wonderful experiences with animals, it is difficult to choose just one to write about. But since this book deals with the urban farm, I will share about my mini nubian goats that not only have a job, but are such sweet babes of the city-sprawl on our farm.

I always dreamed of having goats, with pictures in my head of Heidi in the Alps – a romantic vision, for sure. And true to my dreams, I picked out two females only weeks old, not even weaned from their moms. The larger one I named Amelia and her previous owners assured me she would be a good milk producer. I was also drawn to her cousin, a runt I named Betty. We tenderly gathered up the two and put them in a pet kennel in the back of our small SUV. As my husband and I drove off, the two girls (kids) began to bawl. We were told that they would make it to our home fine without being fed, so I hadn't bothered to bring any milk. I couldn't bear to hear those bleeding hearts any longer, so we pulled over to the nearest Walgreens to buy baby bottles and formula. Like babes, I gathered them in my arms and, with a little help from James, we got them to latch onto the rubber nipples, where they found contentment. With full bellies, they fell asleep on top of each other behind our seats in the dog kennel. Through the bars, I marveled at how perfect Amelia and Betty's formed bodies were, so small and wonderfully made.

Back home on the farm, they did not want out of my sight and cried each night when they went to bed. Due to the cold weather, my husband reluctantly let them stay in our makeshift home rather than the goat pen we spent more on than our own room. I bottle fed them on our bed until they got terrible diarrhea from the goat formula. The price of raw goat's milk was over the top, and my farm friend who raised goats told me that raw cow's milk would do the job fine. So we added probiotics to cow's milk to aid with digestion, and like Pepto Bismol, it worked and their stools began to take the form of bunny droppings!

They imprinted upon me and tailgated my every move, which meant I couldn't leave them. I created children, really, and so when I went out places, I put them in the carrier kennel and off we went. It was so fun to have two girls and we got lots of attention, as you can imagine. At cafes, we had coffee outside, and we took walks through the neighborhood. When I made dinner, they joined me in the kitchen, but then as if overnight, suddenly not only did their hooves touch the table, but now they jumped aboard, grew too big for the car carrier, and their poop got bigger, which made outdoor cafes not such a great idea.

So, like a mommy eagle pushing her eaglets out of the nest, I pushed the kids out of the kitchen and made them stay in their goat pen. They cried at first, but over time they became comfortable with our routine of morning attention and evening walks. Amelia took on the role of mother and Betty followed her lead. When they were bred, they cried when we left them, but came out of that experience smelling of goat. The birth of their kids was shared with one of our dearest friends and is stored in my heart amongst my most special memories.

Though there are several reasons to keep milk goats, for me, the main draw is my love for God's creatures. But did I mention that fresh goat's milk is the best? And no, it does not taste goaty, unless you keep males nearby.

great times raising our egg layers. I learned later as my flocks grew to stop naming them all. The first three years we had plenty of eggs to eat and share, then in the fifth year most stopped laying altogether. We now had a bunch of old hens with lots of mouths to feed. Given that I was so attached to them, I didn't have the heart to eat them or let anyone else, so they got to live their lives out in peace and moved with us to the farm we currently live and work from. Once a vegetarian, it took me a while to eat my own raised meat, but over time, somehow it worked out for our family and others who valued good pasture-raised organic meat. It may help with eating meat to make it into a side dish so you don't need to feel so gluttonous about taking a life, but instead can feel honored to have access to humanely and ethically raised meat.

Chickens, especially bantams, love to be mommies. Set up boxes so the hens can brood and hatch young. A newly hatched chick brings great joy for all ages! In our farm school, many of the young ones' favorite lesson in animal husbandry involves picking red wigglers out of the vermiculture bins and then letting the chicks snatch them from their tiny, previously timid worm-catching hands.

I started my animal menagerie with chickens and then, once I felt confident, I introduced ducks. At first, I was really confused because our alpha male rooster kept mounting my female duck and I thought he was trying to kill her! But then it dawned on me that he was trying to mate. I separated them and my Khaki Campbell ducks were much happier. So happy that, even though their breed rarely hatches, nature found a way and I marveled as, on the day before Mother's Day, out from under a large rose bush emerged over a dozen adorable ducklings waddling behind Susan, their mom.

I have also raised turkeys, but decided I just do not have enough pasture for them to roam, plus I get too attached to those sweet birds. However, if you do have a lot of pasture, heirloom turkeys are in need of being raised in order to keep the breed alive. We buy our heirloom turkey from a genuine farmer out of Grandview. With plenty of pasture, it is very easy to grow the seeded mix they should eat, and is better on the environment than the traditional industrial farming methods using stacked, enclosed turkey houses. Does it cost more money? Of course, because our government doesn't subsidize non-GMO farming. The choice is yours, so if you want a farm-raised turkey for the holidays, then raise one. Before you start though, pick about a quarter of an acre and grow a turkey feed mix. You may have to buy some of their feed in the winter, but it's worth the love.

My sweet goose, Lady

Lady, our lovely Chinese goose who passed this last year, gave many visitors and myself great pleasure and the best deviled eggs anyone ever ate. She is missed. I believe that while I was in the hospital battling my own adversaries, she died of a broken heart, since I was her mate. She came to me after being hand-raised by another woman, who passed away. The son was going to have goose stew unless he could find someone to take her, so on Valentine's Day she came to me as a gift from a friend who rescued her from the pot. While she was with us, she would come sit on our laps and would honk every time you came nearby if she was penned, making it known she needed affection. She loved to come into the open market area when customers were here and let them know who was the boss. She had her favorite customers and then others, well, let's just say we had to pen her while they lingered. Like any lady, she had her disdain for certain people, but overall she was the tamest goose anyone had come in contact with. Lady was one-of-a-kind and is missed, so much. Her last days do have a silver lining. A sweet-spirited young intern loved by lady held her on the last day of her life. She laid her long neck and head down on Mary Katherine's chest and peacefully let her spirit join heaven, where she can now finally sit on eggs that hatch – our chickens often had to fight her off their eggs, as she was created to be a mommy.

RAISING ANIMALS SUSTAINABLY: A REALITY CHECK

I am intentionally being redundant to drive home that in choosing to to raise animals, whether flocks or herds, you need to think beyond three years of fresh eggs or several years of fresh milk. Because on a sustainable farm, every animal should have a job and not just be a mouth to feed, even if your farm is your backyard. Don't get me wrong; bringing animals into the mix increases yield of produce crops, cuts down on pest, increases your compost supply, and if you choose goats, there are several functions they can serve beyond the barn. They are the best hedge and tree trimmers and simultaneously cultivate the ground with their hooves and deposit ready-made organic fertilizer. Besides those functions, they are amazing creatures and well worth the investment of time if you have it to give. I researched goats for five years before we took the plunge. I wanted milk goats and I know that meant a full-time commitment. Nevertheless, sometimes those same animals we love must turn into someone's dinner. And because of this, keeping stock is a serious decision.

I don't take lightly the harvesting of a beloved animal because it is a life we should be grateful for and that sacrificed its flesh to feed us – their friend and also their foe. So, when taking on this wonderful farm practice of raising stock, please consider beyond three years of fresh eggs. What about the rooster factor? What if your milk goats have male kids? Keep in mind that males are much more difficult to find homes for. If you like to run away on a whim, who will tend the farm? Did I mention that farmers don't vacation much? The diehards I know consider farming their life's work and are quite content to enjoy what they have chosen: a passion that gives more than it takes.

Journal Space

Garden Plans

FALL

October

I cannot endure to waste anything as precious as autumn sunshine by staying in the house. So I spend almost all the daylight hours in the open air.

— Nathaniel Hawthorne

OVERVIEW

This is one of the best months in North Texas to plant fruit trees, fruit bearing shrubs, edible vines, and many herbs and roses. You may have opted to let your body and beds rest this fall and winter, so this is a great time to find some good fruit stock, vines, and bushes. All you wildflower lovers should take advantage of October's uncanny ability to season and germinate flowers like bluebonnets, coneflowers, and poppies. The cool is here at last and we are grateful for perfect weather. Have a garden party!

OCTOBER CHECKLIST

☐ Plant winter garden crops

☐ Plant or transplant fruit-bearing trees, shrubs, and vines

☐ Start planting garlic

☐ Build cold frames to use as early as late October

☐ Start a compost pile for spring

1. FALL-INTO-SPRING CROPS: WINTER HARVESTING IN THE CITY

We have a unique situation here in North Texas. Even though we have cold spells, we can grow food year-round provided that some simple guidelines outlined in this book are practiced. This can also be applied in several other regions in the world that have short or mild winters. When I say mild, I am not talking about weather that is above freezing, but rather winters with warm spells, temperatures that can rise to about 60°F, and warming earth, rocks, water, and hard surrounding surfaces.

My neighbors used to plant their broccoli, which requires soils above 60°F to germinate but doesn't tolerate much heat, in the spring months. North Texas, where I do a lot of growing, is a hard place to get a decent crop in the early spring, as in other regions where the weather is shifting from hot to cool. Plants just are not happy then, and before long, here comes the heat, bugs, and more challenges. So my neighbors have decided that growing cruciferous plants is not worth the trouble. Then in the winter, those same neighbors will come to my CSA and wonder how I am growing broccoli in the winter, when by now the weather may be in the low twenties.

I explain to them that I started to sow seeds in early October, once the nights started to cool. Then in late October, I placed cold frames over them, which created something of a greenhouse effect and a fabricated microclimate. This will work in many states that have hot dry summers and unpredictable winters. So rather than focus on spring as my major planting season, I have shifted my focus to cool-season winter crops, which should be started in the early spring in states like Oregon and New York, but that can be grown from the fall into the early spring here in North Texas. Since fall days are shorter and the sun is lower, crops do take longer to grow, but the tradeoff is that more areas in front and backyards are exposed to the sun, whereas in the summer months, they are shaded. This itself opens up a whole new opportunity for homeowners whose houses are covered with the shade of deciduous trees in the summer; they can mimic a food forest by planting lots of cold-season crops beneath the bare canopy of the trees!

Inside a cold frame made with PVC pipe, plants happily grow in the cooler temperatures. The plastic covering can be removed on warmer days.

So in December I have been eating and sharing extremely tasty and juicy cool-season foods such as bok choy, red cabbage, kale, broccoli, and lettuce. Even before my hybrids produce large crowns and heads, I have been enjoying lots of tender and sweet leafy greens. It makes sense that the leaves and roots of my edibles would be so flavorful, since in the winter months, water and natural rainfall are more plentiful and the sun's rays are less intense.

Food storage can be understood as living and growing in the winter and does not require that you do a big harvest to keep fresh produce on hand. In Washington State, where I grew lots of food during the growing season, we harvested and then stored root vegetables in a cool dark place, or canned them. And, though keeping food like potatoes and onions is common, in Texas' warm winter, they sprout rather quickly, unless one has a root cellar. So the problem of having food ready for harvest and storage at all times of year in a climate such as ours is really the solution: grow year-round and let the earth store your food.

Since I am originally from a region where food grows practically overnight due to an intense growing season, it took a while to shift my gardening habits and learn to see my climate problem as also my solution.

I have found winter gardening is my answer to avoiding the warm temperature bugs, drought, and water restriction that are common in our summer months, as well as reducing my exposure to orange alert pollution conditions. I am not saying that I don't grow tomatoes, zucchini, and basil, but rather by recognizing my love of gardening in a bug- and heat-free environment, I have found my solution: winter gardening. I can have broccoli, cabbage and many other favorites without fighting warm-season pests. So again, rather than plant my broccoli in March, I now plant it in October to take advantage of the more favorable growing conditions.

This nifty frame is made of pallets, rebar to secure piping, PVC pipe, vapor barrier rolled off to the side, and the chicken wire is needed if you let flocks saunter.

Cool-season gardening in warmer climates does, however, pose the challenge of protecting plants during the long, hard, freezing days. But just like having pests and other garden enemies, this condition also has solutions, some of them being more time- and material-restrictive than others. One option is to create a guild of plants that can protect and nurse one another. But perhaps the most economical solution, as well as being effective in keeping the foraging chickens out, is to plant pallet gardens with cold frames made of PVC.

Growing in the winter is not a foreign concept for greenhouse growers, most of whom also have favorable conditions in the warm growing season months. With winter temperature control, they are able to produce greater yield year-round, but often at a greater price, due to fuel or electric costs.

Bear in mind, the intent of this book is to bring success to frustrated gardeners in less than optimal conditions. I encounter folks all the time who have moved to Texas only to discover that they cannot garden and some of which have thrown away the trowel. It is a satisfying feeling for these same individuals to come back grinning from ear to ear, telling of their garden success! All gardeners can use these tried and proven principles to increase yield in the garden, while reducing their workload, if they so choose!

EASY RECAP FOR SUCCESSFUL COOL-SEASON GROWING:

✦ All-day, full sun exposure is best, since the days are shorter and colder.

✦ Buffer your crops from being chilled by the northwest prevailing winds.

✦ Plant your garden plot with good compost and have straw, mulch, or thick plastic on hand to cover your crops, should a sudden snap come when your plants are young.

✦ If you have time and resources, build some cold frames using recycled materials.

✦ Pay attention to the weather, since our winters can change drastically, especially in the valley regions and prairie landscapes.

✦ Water well before a cold snap, avoiding too much over-spray into the street, for the safety of car travelers.

✦ Make use of concrete, humidity, plant screening, edges (places that hold onto energy), and all other opportunities to offer warmth to your crops.

✦ Crops planted in the fall will grow well up until December, then will slow down considerably and possibly even become dormant. If you started your seeds as early as September and in some cases October, in the winter you will be eating crowns of broccoli, scrumptious cabbage, roasted carrots, and fresh salads mixed with spicy arugula and green onions. But resist the urge to eat all of your winter crops because those left in until March will max out and grow to be four times the size that they were in the cold winter.

2. WHAT TO PLANT?

If you are planting seeds, make sure to get them in the ground during the first part of October. Remember that the days are getting shorter, so your crops needs as much daylight as they can get to reach maturity.

SEEDS

✦ Carrots

✦ Parsnips — These are tricky, but I like to grow a few a year.

✦ Turnips

✦ Radishes

✦ Peas

✦ Beets

✦ Garlic (cloves)

✦ Spinach — Try to find cultivars that are fashioned for hot days.

✦ Broccoli rabe (or starts)

✦ Salad mixes (or starts)

✦ Kale (or starts)

STARTS

- Rutabaga
- Collards greens
- Kohlrabi
- Onions (sets)
- Broccoli
- Cabbage
- Chard

Basically, the crops to plant this month are all the things you thought you should have planted in the spring! October really is the best time to garden in North Texas, specifically the middle of October. Enjoy and plant, plant, plant!

FRUIT TREES

Besides the common varieties that you have already heard of, there are many food-bearing trees and vines that are not as popular, but do well and thrive in severe climates, given proper care in regard to soil amending and water harvesting. For this reason, I have named a few that are worth the hunt and will add diversity to your farm and garden.

- Black Walnut
- Mexican Plum
- Santa Cruz, Methley, Bruce Plums for hard climates
- Anna Apple
- Pink Lady Apple
- Pawpaw
- Peento Peach
- Red Skin Peach
- Asian Pear
- Mulberry
- Fig – Make sure you get a variety that is cold-hardy and protect it from north and northwest exposure.
- Russian Pomegranate
- Generic Pomegranate works well unless it gets into the teens
- Quince
- Olive (Olea Europaea) Cold-hardy to 10°F
- Loquat

SHRUBS

- Seaberry
- Silverberry Elaeagnus – also known as Tutti Frutti, fixes nitrogen
- Goji berry
- Pineapple guava – a shrub-like tree

VINES

✦ Hardy Kiwi – Male and female. In the year of planting, you will have to baby the male species through the summer.

✦ Anna Kiwi

✦ Cinnamon vine

✦ Grape hybrids – Mars, Thompson, Muscadines, and Champagne

✦ Mustang Grape – a wild cultivar

✦ Passion Vine

✦ Hops – Willamette, Nugget, and Cascade

✦ Blue Passion Vine (Passiflora Caerulea)

✦ Blackberries – Brazos, Cheyenne, and Cherokee. Make sure to get the thorned varieties if you want lots of berries.

Thompson seedless grapes growing at the garden's edge

3. GARLIC

In permaculture circles, we often talk about stacking functions, including the individual functions of a plant. Garlic is among one of those multifunctional plants and is a must for every gardener, whether they like the taste or despise it.

In the fall, I plant red-skinned creoles because I love how they taste and how they perform in hot soil conditions. Other reliable varieties to plant in severe soils and climates are china dawn, silver rose, Spanish roja, and inchelium, which is claimed to be the oldest strain of garlic in North America.

Garlic planted in the fall (late October through December) will be ready for harvest in June. In most cases, you will not need to water it, since garlic is drought-loving and the fall rainwater is plenty enough for a dry crop.

Planting garlic in Greenville, Texas

The beauty of this plant is that while it grows so easily and remediates soils from harmful fungi, it also prepares the earth for late spring crops that are susceptible to blights. Before I harvest my garlic, I often plant tomato starts between the rows. Garlic is a natural fungicide and a great companion to most any plant.

OTHER USEFUL GARLIC REMEDIES, BACKED BY SOUND SCIENCE AND LIFE EXPERIENCE

+ Garlic tea keeps the mosquitoes away.

+ Garlic is beautiful in the garden.

+ Garlic lowers blood pressure (9% to 15 % with one or two medium cloves per day).

+ Garlic lowers LDL cholesterol (9% to 15 % with one or two medium cloves per day).

+ Garlic helps reduce (plaque) within the arterial system.

+ Garlic lowers and helps to regulate blood sugar.

+ Garlic helps to prevent blood clots from forming, reducing the possibility of strokes and thromboses (hemophiliacs shouldn't use garlic).

+ Garlic helps to prevent cancer, especially of the digestive system, prevents certain tumors from growing larger, and reduces the size of certain tumors.

+ Garlic may help to remove metals in the human body such as lead and mercury.

+ Garlic has antifungal and antiviral properties, not only for the human body but on plants as well.

+ Garlic dramatically reduces yeast infections from Candida species.

+ Garlic has antioxidant properties and is a source of selenium.

+ Garlic tastes great, no matter how you crush it!

Garlic benefits from Keene Organics: http://www.keeneorganics.com/hebeofga1.html

4. TRANSPLANTING FRUIT TREES AND OTHER PERENNIAL CROPS

In North Texas, October is the best month to do most gardening! The heat has waned, the sun is still warm, but the soil in most cases is about 75°F at the root depth, and the plants are happy. So they come out after the summer heat for a season of song and dance! It is a shame that most Americans get boxed into thinking that spring is the garden's utopia because it is just not so in many hard and difficult landscapes, North Texas to name one. In the prairie, there is no better time to garden than the months between the fall equinox and the winter solstice –

ah, my anniversary and birthday!

So enjoy gardening year-round, especially when the earth is clothed in rich vibrant hues of red, gold and brown. The cooler weather of October will help to lessen the shock of transplants and allow your prized boys the chance to grow a bit more before nature goes into Winter's sleep or, in a lot of cases, several cat naps.

To transplant a tree, you will need compost, a natural liquid fertilizer for soaking, and mulch for top dressing and to create a tree well.

Make sure to prepare the new planting area first before you dig up your tree!

When digging up your tree, keep the tree's root ball intact as best as you can. If the roots have to be pulled apart or cut to bring them up from the earth, soak your tree for an hour in liquid seaweed or compost tea.

When installing, never put the base of the tree trunk below ground level. Rather, make the hole three times as wide as the root ball, but not any deeper. Roots grow to the side anyway, unless of course, we are talking about a plant that has a tap root. In that case, the roots will dig deep down with profound strength. If you plant your tree too deep, it will will suffocate and die over time, or rot from poor drainage. This is a very common mistake and I have had the opportunity to save many a tree's life by pulling it up by the neck and packing the hole's bottom with good drain materials before placing the tree in a remodeled home.

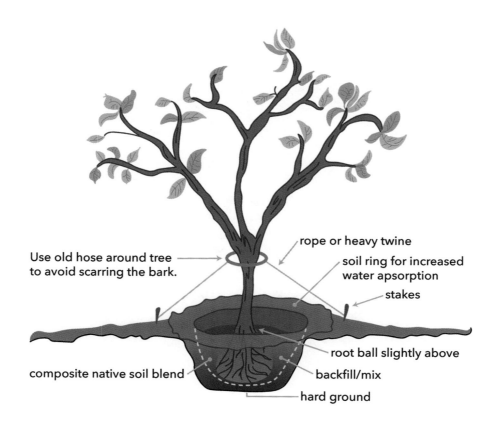

Use old hose around tree to avoid scarring the bark.

rope or heavy twine

soil ring for increased water apsorption

stakes

root ball slightly above

composite native soil blend

backfill/mix

hard ground

HOW TO PLANT A TREE FROM A CONTAINER, FOR BEST RESULTS

1. Dig down the same depth as the root ball; <u>no</u> deeper. This is important to keep the tree from sinking, which causes rot and suffocation.

2. Dig the width double the width of the root ball or container size.

3. Use compost and native soil to backfill the width. Use remaining soil for tree ring.

4. Water well!

5. The tree will be rooted into the new ground in 30 days.

Autumn Memories

Memories of cold garden days, wet with dew and frosted leaves laid flat, revealing acorn squash sweetened by October's chill, make me smile. It really is the chill of fall that brings out the sugar in the flesh of winter squashes. Autumn's joy abounds with sweet meat from green vines. But do not plant winter squash in the fall, but rather in late spring, and then nurse them in the heat of summer with shade cloth or an intentional guild. Spray them with neem oil or wash with diluted orange oil to keep bugs away, and try growing them vertically to best utilize space and also to enhance air circulation.

5. A FRIENDLY REMINDER ABOUT COMPOST

Now is a great time to start gathering leaves for your spring compost. We like to take our truck and trailer out on leaf pick-up day to make a haul. The city is such a great place to collect leaves and you don't even have to rake and bag. The only downfall of using bagged leaves is that you will need to filter them for pieces of string and trash so that your flocks do not get tangled up. We lost a hen this way, as she fatefully hung herself in a large rose bush. Did I mention there are casualties that can be avoided but still do happen?

It takes about six months for leaf matter to break down into compost, assuming you keep it moist and add some green nitrogen sources. Besides manure, coffee grounds are another free and great material that will heat up your pile to become a valuable resource for next season's planting – compost!

Journal Space

Though I don't talk about artichokes in this book, as they are more challenging to take care of, they can in fact be grown in our climate! Try them out once you have mastered the basics.

November

They will come and shout for joy on the height of Zion, and they will be radiant over the bounty of the LORD — over the grain and the new wine and the oil, and over the young of the flock and the herd; and their life will be like a watered garden, and they will never languish again.

—Jeremiah 31:12

OVERVIEW

What a great month to enjoy the fruits of your labor and to plan your holiday meal around the crops you have grown. If you have greenhouses and cold frames, you will need to tend to these crops; otherwise, the leaves will lay down to rest on the earth's floor. Enjoy free mulch (leaves) and water the soil if it gets dry.

I have intentionally kept this month short and simple, since in the world of growing food, I need to take a respite and tend to a plethora of other activities and events. I love that about gardening the unconventional way, that once you have systems in place, they can be neglected and managed by nature. If you have livestock, however, they may need more attention, due to the cold weather. This is why shelter is high on the needs list before you get your flocks, hives, and herds.

> **NOVEMBER CHECKLIST**
>
> ☐ Finish winter planting, if you haven't already
>
> ☐ Tend to winter beds that have cold frames
>
> ☐ Water the soil and fallen leaves if it gets dry
>
> ☐ Cook a Thanksgiving feast!

1. WHAT TO PLANT?

I love to garden in November, since it is mosquito-free and many other pests such as aphids and white fly are at bay. I have usually planted the crops that fair better by seed in October, since the days are already getting shorter and the soils are cooling below germination temperatures. But there are still several starts and seeds that you can plant as late as November and harvest in late winter and early spring. These are the plants I have success with. They can be tough to the cold and require little care other than some hydration.

SEEDS

+ Dockweed (or starts)
+ Carrots
+ Radishes
+ Beets
+ Borage (broadcasted)
+ Turnips
+ Garlic (cloves)
+ Onions (sets)
+ Leeks (sets)
+ Asparagus (crowns)

STARTS

+ Kale (all varieties)
+ Mustard
+ Arugula
+ Collards
+ Parsley
+ Cabbage, red and white
+ Hardy wild lettuce
+ Broccoli

Did you know that the whole carrot is edible? The greens taste like carrots and are yummy when they are young. Carrot greens make a tasty garnish and are good for you! These are usually plentiful in the late winter, if you plant them by seed in October.

ESTABLISHED PLANTS

+ Culinary Sage (4-6" plants)
+ Rosemary
+ Strawberries
+ Rhubarb
+ Blackberries

+ Kiwi Vine
+ Haskap Berries
+ Grapes
+ Most fruit trees (see October list)

2. THANKSGIVING FOODS, ETHICS, AND INSPIRATION

If you know your way around with keeping chickens, consider raising some turkey hens.

But beware: these cultivated varieties are meant for the table, and thus do not have a long life span. I had to learn this lesson the hard way, and my attachments proved I had some growing to do before I could eat my hand-raised flock. Turkeys are extremely sweet-natured.

One year, I grew five birds and still ended up buying a local organic bird to eat instead. Though we had to harvest my sweethearts anyway due to their bird variety, I could not bring myself to eat my girls. Instead, they went to other families who couldn't afford pasture-raised turkey!

Now that I've told on myself, let me say this: I was raised around forests, family farms, and ranches, so humane animal slaughter was introduced to me at an early age. I was not a huge fan of meat, and my experience with hand-raising a calf whose mother died during birth sealed the commitment of being a non-meat-eater for me. I was not even ten years old, but my parents, though concerned, consented. (Bear in mind that this was over forty years ago.) And so my mom made sure we had plenty of fish. For some reason, eating fish did not bother me, even though I would catch them and watch their gills bleat for air. My family was entertained by my conviction, at best, and thought it would pass. But by fifteen years of age, I was a serious vegetarian and would remain one until my twenties. By the time I was in my forties, I came to a place in life where I realized I needed to eat meat regularly. This change did not happen because I don't value animals; I do, highly. But after struggling with anemia and depression, I finally had to agree with my body and now my heart is at peace because my body feels better with meat – not commercially grown meat, but meat grown by farmers with values, who respect the connection between plant and animal.

I have even grown to the place where I no longer insult others and, thankfully, eat most meals served to me. If you don't eat meat and feel good, then you are blessed. But please don't think of yourself more highly than you ought. I agree with Michael Pollan: "Eat food. Not too much. Mostly plants."

It is hard for me not to judge the meat industry in this world. But then, who made me the judge? Just do the best job you can do and let the universe and God open the eyes of others. After all, we all have blind spots.

I pray that this book, a mere sketch of my life, leads you into gardening, family, and love – a love that shines light into our dark world and draws some to return to their hunting and gathering roots. Along with my prayer, I am sharing part of my sacred past: my Thanksgiving meal that stretches across borders and waters.

Slow Food

I didn't come from a fast food family. Days of slow food were not intentional, but rather just a rhythm of our lives.

For holiday meals, we gladly gave of our time to make preparations so time together with family and loved ones would be satisfying. I remember my mom spending days beforehand gathering the freshest produce available and staying up late the night before doing all she could to complete all the finishing details that made our home and table a lovely display.

I polished the silver and my father followed behind my mom, cleaning pots and pans as she cooked. Everyone took part in basting the turkey – this may be why we ate hours later than planned every year, since we let the heat out of the oven each time – and my older brother would sneak a fast sample of the succulent meat with each drizzle.

One of the sides I cooked was fresh cranberries. Once you eat the real deal, you will not want canned cranberries again! My mom's called for oranges, which give the garnish a citrus kiss.

But my favorite side dish was one my grandmother taught me to make. Every holiday, though the recipe was written

Winter produce found its way into many dishes.

somewhere, I would ask her again how to make her baked corn dish. The recipe consisted of fresh corn, green onions, a touch of grated carrots, whole milk and cream, sharp cheddar cheese, salt, pepper, and a pinch of nutmeg. It's a very simple dish that keeps me humble – though we all loved gourmet cuisine, when it came time for Thanksgiving, my family preferred a traditional meal.

So as you can imagine, a fresh - not frozen - turkey was ordered days ahead, yeast rolls were baked, pies were ordered from a local pie maker, and even more desserts and salads were offered from other family members. Generous portions of brussels sprouts, sweet potatoes, mashed potatoes with plenty of milk and real butter, and herbed bread stuffing with homemade gravy were the staples of our family's table.

Getting everything ready at once was like a production: my dad mashing potatoes, my mom whisking the gravy, and my grandmother directing the young ones to help or get out of the way, all the while bossing my grandfather to light the fire or dim the lights. Somehow despite all the clatter, the grand finale came when we sat together around the table while my grandfather said the blessing. He was honored to do it and I don't remember a time when he didn't shed tears of thankfulness for his family. This act of gratitude has kept with me all these years and I have passed it on, as we still take time in our growing families to thank God for his goodness.

Along with the delicious food, my mom put out her classic china, crystal, crisp white napkins, and fresh flowers. Nothing was forgotten. Even as children, we were allowed a small taste of good wine, but the best was saved for the adults.

The only burden was the dishes, but we all helped some. My dad stayed up late into the night putting up all the food, returning the good flatware, dishes, and goblets back to the celebratory cabinets, and making sure the kitchen sparkled.

A meal, well done, is a living prayer and a vibrant song of worship.

3. A LOCAL HOLIDAY MEAL, FROM MY FARM AND FAMILY TO YOUR HEART AND TABLE

All of these old-fashioned dishes have their beginnings at the farm of my grandparents – and before that, from their families who came over from Ireland and Germany. My mom has added a bit of her culinary zest, then some dishes are mine, with undertones of earthly likings, and we've included one salad from my husband's side of the family. I hope this collection warms and nourishes your whole being and strengthens the traditions shared around your harvest table.

Elizabeth at her mom's house
on Christmas Day

A Texas Thanksgiving Feast

- ✦ Fresh free-range oven- or firepit-roasted turkey, with all the giblets for the dressing
- ✦ Giblet dressing
- ✦ Creamy turkey gravy made from the fat and trimmings of the bird
- ✦ Roasted sweet potatoes, with marshmallows melted on top for the kids
- ✦ Old-fashioned mashed potatoes, with milk and butter for a creamy texture
- ✦ Baked corn casserole
- ✦ Brussels sprouts with pearl onions, in a white sauce made with sherry and a pinch of nutmeg
- ✦ Fresh coleslaw with bay shrimp and grape tomatoes, unsweetened and with mayo and lemon juice
- ✦ Molded peach Jell-O salad
- ✦ Homemade soft yeast rolls
- ✦ Fresh cranberry sauce
- ✦ Pies, and lots of them!
- ✦ Whipped cream and half-and-half
- ✦ Good wine, red and white
- ✦ Brewed coffee
- ✦ And a hearty appetite!

Journal Space

December

In the depth of winter I finally learned there was in me an invisible summer.

— Albert Camus

OVERVIEW

This is a good month to neglect the garden and take stock in yourself. If you are growing winter crops, you will notice that most of your garden is in slow motion. If you have cold frames, you can relax, since your crops are protected. If you have chosen to not grow crops in the winter, you can still use observation as a good excuse to get bundled up and take in the fresh air while you enjoy Winter's wonders. You may want to take your camera or a sketchpad with you.

DECEMBER CHECKLIST

☐ Take care of cold frame beds, or relax and observe

☐ Harvest rose hips

☐ Make holiday gifts

☐ Buy a living Christmas tree

☐ Dream and plan for the new year

1. EDIBLE ROSE HIPS

Joan is an old friend who I met here in Fort Worth when my garden shoppe was in a mall plaza courtyard. My old-fashioned plants and her cottage pastry business were drawn to each other! Joan's stories of gardening in England evoked memories of my own gardens in Washington State. I was familiar with rose hips, but it was not until Joan shared with me that as a child she would go about the countryside gathering hips from both wild and cultivated roses, then sell them to the pharmacist for pocket money, that I realized the value of roses in the garden.

Rose hips offer loads of Vitamin C and the blooms are also healthy to eat. Dutcher, one of my favorite antique roses, often blooms all through December. This rose can be enjoyed through the holidays, with her creamy flowers cut for décor or sprinkled on holiday cakes and cookies.

HARVESTING ROSE HIPS

December is a good time of year to harvest the seed hips. By now, a good frost has set the color and sweetness. I like to wear rose gloves and use my hands to pluck the hips from the thorny bushes and vines. If you over-cut the dead heads in the fall, you will not get as many hips, or they may be green when it is time to harvest.

After you have picked your rose hips, let the orange seed pods dry out naturally. To make a winter tea high in Vitamin C, crush the hip and then place it in hot water. The taste has a hint of apple, which is a cousin to the rose.

ROSES THAT PRODUCE HIPS AND FRAGRANT BLOOMS

This list contains just a few of my favorites. The rose industry is huge and, while most roses do produce hips, this is a selection of those large enough in size to be worthy of harvest. Rose hips are very versatile and have many culinary and personal care uses. This is good cause to brighten every garden with the plant that Shakespeare thought best.

+ Altissimo – vibrant, red climber

+ Penelope – loose shrub with lovely yellow-pink flowers that fade to white

+ Pat Austin – David Austin tall shrub with petals that cycle from orange to cream

+ Kathryn Morley – David Austin shrub with the most lovely apple and honey fragrance

+ New Dawn – shade-tolerant pale-pink climber

+ Seminole Wind – pink-coral climber that sprawls when left on its own

+ All Rugosa roses

+ Mutabilus – sweet, carefree China bush with a variety of pastel-colored blooms

+ Basye's Blueberry – shrub with blue-purple petals and purple canes in the winter

+ Don Juan – very large pink or red climber and a repeat bloomer

+ Dutcher – creamy white China shrub with a lovely aroma reminiscent of ripe green apples. This is one of my favorites and is worth growing just for its flavor enhancements to rose jelly.

✦ Ginger – stunning upright plant that blooms with hints of pink, coral, and yellow in each bud to bloom. More of a floribunda rose than a tea rose, and one of my favorites because of its unique shade and cabbage head bloom.

The best time to plant roses is from October to March, assuming they are grown outdoors. Greenhouse roses do the best from October to April. But overall, roses are tough girls and can be planted year-round as long as you prepare your soil and protect the tender plants from freezes. If you are planting in the heat of summer, water your roses daily for the first thirty days, then gradually reduce to a regular watering schedule. More roses die from over-watering than from under-watering, so they are a good choice for the edible Texas garden.

Roses take the cake!

Early in the morning, before even a creature was stirring – not even my mom – I tiptoed out into her freezing West Coast Washington garden and planted a rose, the Belinda's Dream that I had smuggled onto the plane in my carry-on bag. It was December, so by now, my lady was toughened up to some cold weather. In my flannel pajamas, I dug a nice hole in the wonderful soil, in a spot just calling for a cabbage-head rose, near some other roses that were laid to rest for the cold winter. The shovel broke easily through the light freeze on the ground and, in a dash, I had planted the promise of spring for my mom. I had hoped to later surprise her completely, but like any intuitive gardener, she noticed the now leafless Belinda's Dream not long after, asleep in her dirt bed, and suspected I had been up to my elbows in the freezing cold to put it there. And so the story goes that it was an early surprise, and again an early spring surprise, for my mom to be greeted by one of my tried-and-true rose choices. All this to say that roses are tough girls and have the thorns to prove it. Don't be afraid to grow these cousins of wild berries, but rather, be bold and plant in good dirt!

2. WINTER-HARDY HERBS

The sun makes a tomato taste like summer in the same way that the frost brightens the flavor of cold-weather evergreen herbs. No garden seems complete without the pungent fragrances of rosemary, lavender, sage, and thyme.

Though there are other herbs that will grow through the winter's coldest months, these are evergreen and are a mainstay for seasoning savory holiday meals. Rosemary, well known for its many uses in the kitchen and medicine cabinet, is also an herb of remembrance that is easy to slip into Christmas cards or use to garnish a gift.

A December hill of greens

3. WINTER REFRESHING

Baths infused with homemade lavender are in order after stressful and cold days. An easy and low-mess method is to pack some cheesecloth full of fresh lavender, whether in bloom or not. Then run really hot water through the bath bag to extract the oils, or make a concentrated tea the same way and add it to your bath. Not only will you feel great and refreshed, but the aroma is like therapy for your home.

If you are not a bath person, build a contained outdoor fire and burn dried herbs with cedar for an aroma sensation. Woody herbs such as sages and rosemary work best for this. When harvesting and preserving your herbs for winter, hang them in a cool, dry, dark place. After they have dried, remove the leaves from the stems, saving the stems to use later as fire starters with the same effect. Glass jars with lids come in handy for storing dried herbs. Label and tie ribbons around the jars, then give as holiday gifts to all the foodies in your life.

4. LIVING CHRISTMAS TREES

It is admirable for folks to go the extra effort of buying a beautiful potted evergreen tree to bring the Christmas spirit alive. The climate here doesn't provide proper conditions for traditional firs, due to our hot summers. I have found a few good alternatives though, and all the trees are lovely, really. When selecting a living tree, pay attention not just to the variety, but also to the transition of the tree from living outdoors to indoors. The longer you can wait to bring your tree inside, the better, and then keep it away from heat sources. Make sure your tree gets plenty of rain or filtered water. Limit the amount of hours you have the lights turned on and use low voltage lights. Once Christmas week is over, do not move your tree directly outdoors, but rather transition it to a garage or shed. You get the idea: be gentle with the transition, then after a couple weeks, start a new tradition of planting your living tree.

The following trees are my favorites, thus far:

- ✦ Arizona Cypress – drought tolerant

- ✦ Deodar Cedar – lovely soft evergreen with a weeping shape

- ✦ Blue Point Juniper – bushy cone-shaped evergreen

- ✦ Leyland Cypress – though stunning, does not do well in drought; so if you choose this route, go the extra mile and dig an extra large and wide tree hole, fill with a compost and grow-core mixture, then use gator bags for the first three years.

- ✦ Citrus Trees – break the mold! But don't take outdoors until spring.

5. DESIGN – LET YOUR IMAGINATION COME OUT AND PLAY!

The gifts are put away, the guests have gone home, the kids are content with their new toys, and Dad is back at work. Now you may start seeing visions of tomatoes, lettuce, and cucumbers bursting forth, and are even entertaining the thought of planting a small food forest or an edge bed along your back fence!

To make your dreams a reality, first you will need to start with notes and a site plan. Though creating a site plan is fairly simple, you will probably find one filed away from the purchase of your home. Even if you cannot find yours, if you live within the city limits, your site plan (commonly called a "plot") should be available through the plans and development department of your city. The advantage of a drawn lot from the city is that the measurements will be accurate, since they were created by a professional surveyor. With site plan in hand, your job can be as easy as making copies and filling in the page, or you can use the measurements to make a scale drawing on drafting paper. I prefer the latter so that I can blow up the lot size to scale and have plenty of paper space for drawing in each plant, tree, path, structure, and feature, along with notations.

Winter Solstice Prayer

The winter solstice occurs on December 21st to the 23rd, the specific day dependent on which year this astronomical event takes place. The following is for all you holiday-born babes who have felt overlooked: the winter solstice is not just the shortest day of the year, but also the breaking of darkness into the coming spring light.

I tend to think that winter babies grow to be more grounded and intuitive people. My friends born in the warm, fertile months have an edge on the rhythms of nature also, but gardening comes especially with ease to those born in the winter solstice's stillness of time.

Snow chickens!

I pray for snow, Winter's throw,

So I can lay across my efforts; farm work buried, I breath deep.

Now the soil, weeds, and clover are content to rest.

Daybreak is just weeks away; the winter-into-spring light will soothe my soul.

Journal Space

Garden Plans